A Practical Guide to

Alterations and Improvements

by

Ron Cooper

and

John Buckland

International Thomson Publishing Limited,
London WC1X 0BP.

Published 1985

ISBN 1 85032 006 3
A 'Building Trades Journal' Book

Typeset by Avon Communications Ltd, The Sion, Crown Glass Place, Nailsea, Bristol BS19 2EP

Printed and bound in Great Britain by John Wright & Sons Ltd, at The Stonebridge Press, Bristol

CONTENTS

Introduction

This series of projects is aimed at helping the small builder to estimate and tackle with confidence the everyday types of improvement and alteration jobs. It is hoped that any pitfalls likely to occur can be foreseen so that the reader will learn by these projects and not by mistakes during the course of the work.

The projects do not explain in infinite detail all aspects of the work involved as some knowledge of the building function is presumed.

On the pricing information given a standard rate of £4.60 per hour has been used throughout whether for tradesman or labourer. The nature of the work dealt with in this book is such that the role of the tradesman or labourer is frequently reversed as needs require. Also, on very small works a handyman/tradesman is often used.

A general overhead and profit margin of 40 per cent has been allowed throughout and this must be considered very much an arbitrary figure as it will vary considerably according to the size and general organisation of the company involved. Also, profit levels on very small jobs need to be higher than on larger contracts.

In every case there are additional costs not necessarily covered in the rates or overhead/profit margin. These have been allowed for as prelims at a set percentage of 12½. It is considered this allows for non-working site supervision, small tools, transport and sundry expenses.

Project 1
Double hung sash windows

The majority of house owners are inclined to indiscrimately replace these very practical and attractive windows with their modern day counterparts of aluminium or standard mass produced side hung casements. This is usually thought to be a cheaper and more efficient form of replacement, although such is not always the case.

The builder invited to quote for this type of job will have to take much more into account that just the price of a purpose made window and his labour costs. Scaffolding (either traditional or tower) is an essential and costly item and should be erected and maintained in strict accordance with the Construction Regulations

(Working Places). If the scaffold projects over a public highway, e.g. the pavement, then further precautions and lighting have to be provided.

The builder in this instance will be legally known as the main contractor and so is responsible for all aspects of the job, including scaffolding. Building Regulations approval is usually not necessary for this type of work unless the opening is structurally altered or the areas of openable window are reduced.

Working in occupied premises always involves hidden costs and a percentage is usually added to the estimate to allow for this. The employer's bedroom cannot be treated as a building site! Removal and protection of furniture, curtains, etc., cleanliness and security are of great importance and the owner of the house will probably be equally concerned about the way in which he and his property are treated as to the quality of the job itself.

Double hung sash windows were traditionally made with a cased frame as shown in Fig. 1 and the weight of the sashes is offset by cast iron weights in the frame. When the new window is constructed, the sashes will probably be of a different weight to the existing, and so new counterbalance weights should be provided. Sash chains should be used where the size and weight of the sash would place too much strain on the cord.

Some of the larger joinery manufacturers sell a range of factory made double hung sash windows and though these are normally

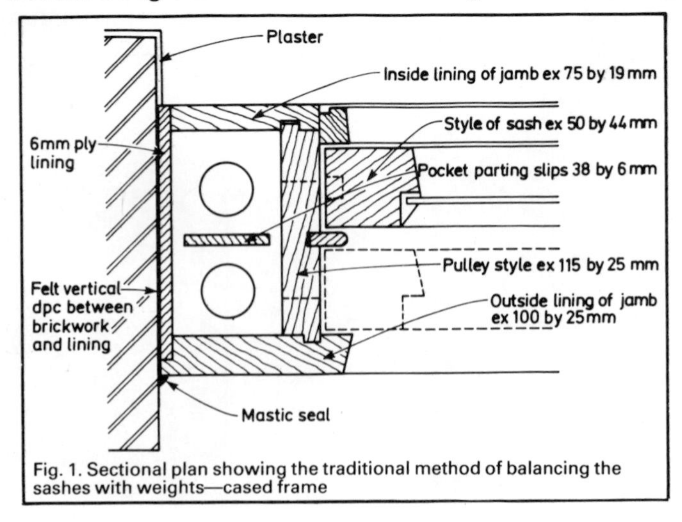

Fig. 1. Sectional plan showing the traditional method of balancing the sashes with weights—cased frame

of an acceptable standard, their sizes and styles may not match the original windows of a house. It is normal practice today for these sashes to be operated by a system of spiral springs instead of cords, pulleys and weights. Speed of production and fewer materials mean that costs are subsequently reduced. This type of construction relies upon friction holding the sash in place and although it is effective, the appearance of the frame is altered considerably; see Fig. 2.

The construction, design and quality of double hung sash windows is defined in the current British Standard. The small builder is unlikely to have the resources to produce the window himself and there are many competent joinery workshops around that specialise in the faithful reproduction of all types of windows. Redwood is the most commonly used timber for the construction of windows today, although if the builder has priced for an exact duplicate the existing type of wood should be determined. Oak, teak and pine were also used extensively in previous years and the price could vary considerably. The sill, however, should always be constructed of a more durable wood like oak.

After scaffold is erected and preparations are made to protect the houseowner's belongings, plastering should be hacked off internally and rendering to the external reveals removed to allow the old window to be taken out. If the stone sill is damaged then this too should be carefully replaced.

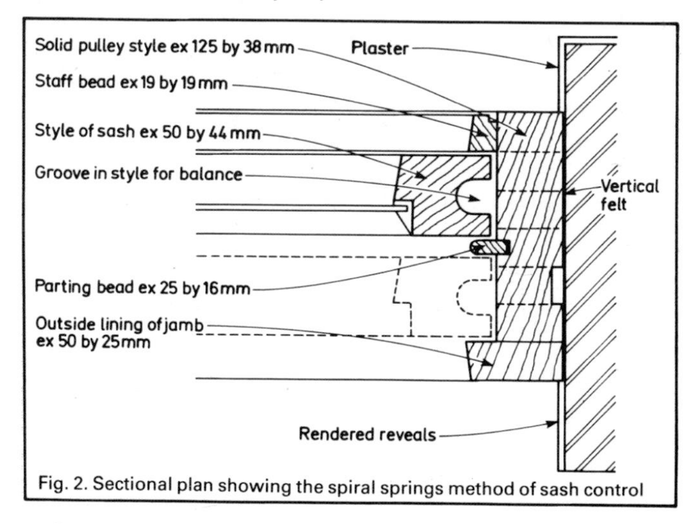

Fig. 2. Sectional plan showing the spiral springs method of sash control

Repairs to brickwork and pointing should be taken into account and although this may be an unforeseeable item at the time of estimating, allowances should nevertheless be made.

Fig. 3 shows the relationship between sills, window and brickwork. Throating and weathering to both the timber and stone/concrete sills to properly throw off rainwater and avoid dampness are necessary if the window is to last for any length of time. You only get what you pay for, and this is especially true of joinery. It is not worth trying to cut corners and use inferior or smaller section timber as the shorter life of the window will only reflect this false

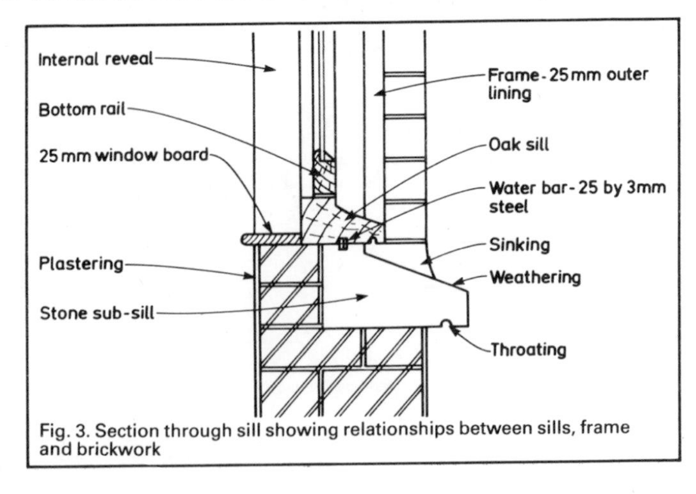

Fig. 3. Section through sill showing relationships between sills, frame and brickwork

economy in the long term. The importance of careful preparation and regular maintenance should be stressed to the client at the time of the installation.

The correct size and style of sill may not be readily available and so a reconstituted stone or cast concrete sill will have to be made up. A water bar should be inserted between the top of the stone sill and the underside of the oak sill. This water bar is bedded in mastic in the oak sill and in cement in the stone sub sill and acts as a barrier to the entry of water between the two sills.

When fixing the new window, either frame cramps fixed into the brickwork or plugs and screws can be used. Prior to offering it up, however, it is important to fit a section of the felt vertically to either side of the opening to alleviate the possibility of dampness affecting the unseen parts of the timber frame. Care should be taken though not to damage the felt when fixing the frame.

Remember that the horns of the frame should be removed and the ends properly treated before the window is positioned. It also goes without saying that any other areas of timber that will be inaccessible once the window is in place should also be treated with a couple of coats of a good quality primer.

Ensure that the window is level and properly positioned before fixing. Renew the internal window board and make good to internal plaster and any external rendering. When glazing the windows, use wooden sprigs and putty and do not forget that the sash weights will have to compensate for the weight of the glass as well as the sashes themselves.

On high class work, hardwood glazing beads and washleather may be used to glaze the windows, as long as this matches the other sashes around the property.

If there is no external rendering, a mastic sealant has to be run around the edges of the frame as a barrier to water penetration. Internally, fix a softwood architraving around the sides and head of the frame to match existing.

The extent of redecorations will obviously depend upon the job and the client. In all cases though, the window itself should be properly prepared, knotted, primed and stopped and given two undercoats and one top coat of hard gloss oil paint. I cannot stress enough the importance of adequate maintenance and thorough redecoration, especially externally, if the joinery is to last for any length of time.

On completion of the works, clean the glass of both sides and ensure that the frame is clean and tidy internally. Clear away all rubbish, surplus materials, etc. and supervise the removal of scaffolding.

Build-up of estimated cost to replace double hung sash window at first floor level:

When pricing an item such as this it is necessary to consider whether it is a job in isolation, a single window of which there are several similar but in different locations, or an item of work within a much larger contract.

The prices that follow are for one window in isolation:

		£
a)	Clear and protect area in front garden for scaffold tower (one man ¼ day)	12.56
b)	Supply, erect, maintain, dismantle and clear away 2.00 metre square scaffold tower 6.00 metres high including all necessary scaffold boards; transport on and off site,	
	Scaffold hire, say one week	15.00
	Transport	20.00
	One man one day	50.23
c)	Move furniture and protect as necessary (one man ¼ day)	12.56
d)	Hack off internal and external reveals to free sash and clear away (one man ¼ day)	12.56
e)	Remove sash from opening and carefully lower to ground for transport off site (two men ⅛ day)	12.56
f)	Repair or make good reveals to opening including cill and arch as necessary,	
	Two men ⅓ day	33.16
	Material, say	22.00
g)	Allow for weather protection after removal of sash if new sash is not fitted immediately, (provisional allowance)	25.00
h)	Allow for making good pointing around window opening,	
	Say 2 SM—one man ¼ day	12.56
	Material, say	3.50
i)	Supply only new window (PC)	240.00
j)	Prepare new window prior to fixing including all necessary knotting and priming and removal of horns,	
	One man one day	12.56
	Materials, say	2.20
k)	Supply and fix frame cramps and cut pockets in reveals for building in,	
	One man ¼ day	12.56
	Materials, say	3.30

l)	Supply and fix damp course between frame and reveal,	
	One man ½ hour	3.22
	Material, say	2.20
m)	Offer up window to opening, level up and plumb before building in cramps. Bed cramps in cement mortar and replace and re-bed any bricks removed,	
	Two men ⅓ day	33.16
	Material, say	2.20
n)	Point around frame externally in mastic,	
	One man one hour	6.44
	Material, say	5.50
o)	Make good all internal plastering and external rendering to match existing,	
	One man ⅓ day	16.58
	Material, say	4.40
p)	Glaze sashes in appropriate glass with putty and sprigs,	
	Assume 1¼ SM glass	30.00
	One man ¼ day	12.56
q)	Re-balance sashes with correct weights,	
	One man ⅓ day	16.58
	Material, say	13.20
r)	Prepare and paint all exposed timber surfaces to match existing,	
	One man ¾ day	37.67
	Material, say	6.60
s)	Make good all disturbed or damaged decorations internally and externally, *Allow Provisional Sum*	20.00
t)	Clean windows, clear up and remove all rubbish; remove temporary protection, reinstate furniture etc. and leave all clean and tidy to employer's satisfaction,	
	One man ¼ day	12.56
		725.18
u)	Allow additional preliminary costs including supervision, transport, small tools, etc, 12½%	90.65
		815.83

NOTE: The foregoing rates have been based on £4.60 per hour plus 40% for profit and overheads. This would necessarily have to be adjusted if the job consisted merely of replacing one window.

Project 2
Side hung casement windows

Obviously, if the window concerned is above first floor level then a system of scaffolding will be necessary. This was dealt with to some extent in Project 1, but basically the scaffolding should comply with the Construction Regulations (Working Places) and is the total responsibility of the main contractor, in this case the builder carrying out the window replacement. The correct scaffold for the type of work should be constructed.

Officials from the Health and Safety Executive (the old Factory Inspectors) have the power to stop the job if the scaffolding is not erected in strict accordance with the regulations.

A suitable case for treatment.

Many of the items mentioned in Project 1 are also relevant in this instance, especially the part about working in occupied premises. Protection of furniture and belongings and security of the house are of prime concern and should be given the importance that they deserve.

The defective double hung sash window should be removed as carefully as possible from its opening after first being exposed on both sides by hacking off plaster and any external rendering. If there is no external rendering, then it is likely that the brickwork above the opening is supported by a brick-on-edge arch. Figure 4 shows a typical elevation.

Altering the shape and size of the opening is going to mean that the loadings above the existing opening will increase with the wider window. Hence, the existing arch will not be of the correct span to support the load. Figure 4 shows this more clearly.

Either reinforced concrete or Catnic-type steel lintels should be inserted to span the new opening with a bearing of about 150 mm at either end depending on the client's requirements and the external finish of the building: face brickwork or rendering.

The size of lintel and padstones should be checked with the Local Authority's Building Regulations department. In a case such

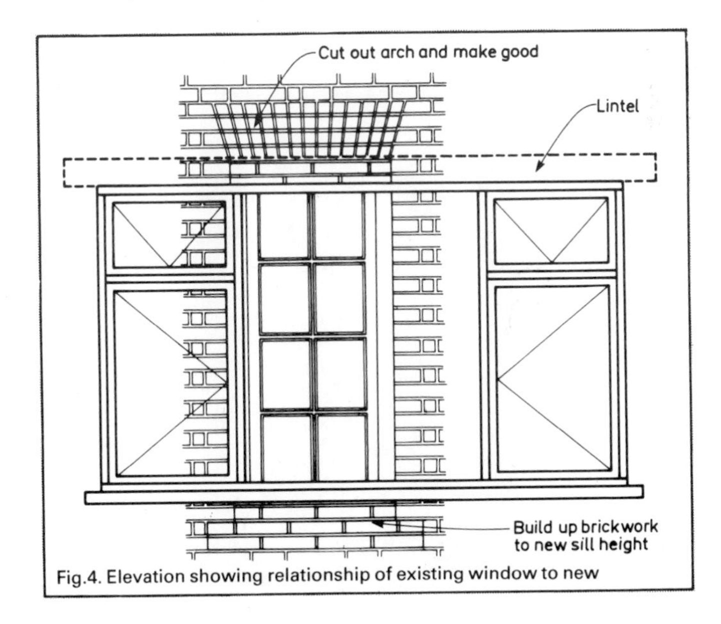

Fig.4. Elevation showing relationship of existing window to new

as this, approval should be obtained under the Building Regulations (or London Building Acts) before any work is carried out. A standard lintel type which meets with the Local Authority's approval should be chosen. Non-standard or complicated locations may need engineer's calculations to prove that they are satisfactory. The calculations will need to be submitted with the application.

Local Authorities now make a small charge for applications and your council offices will send you the scale of fees along with the necessary forms. Make sure to read the small print though and remember that the Building Inspector will want to see the lintel before it is covered over or any making good is done.

If the external elevation is of facing brickwork, then the existing arch will look rather odd perched a couple of feet above the new window. It is best in this instance to allow in the estimate for removing the arch completely and making good above in brickwork to match the existing as far as possible. Aesthetically, a mixture of side hung casements and double hung sashes on the same elevation will not appear correct, and some people would argue that the appearance of an arch no longer supporting anything is no odder than mixing the windows. It is recommended that this type of improvement be carried out only on rear elevations of a house.

If the arch is to be left in place, it is a good idea to hack off some of the plaster internally behind it to expose any timber that might have been inserted to act as a lintel. Poor brickwork and defective pointing to the arch may have allowed the entry of rainwater and rot could have set in.

This type of unforeseeable work should be included for under the heading of a Provisional Sum in the estimate. The client will understand that it is not always possible to price for every conceivable item in a job such as this and the inclusion of provisional sums means that he will not have to pay for something that may not require doing.

Brickwork built in the Victorian era is prone to become porous. One way of solving the problems that this can because is to apply a couple of coats of render to the elevation concerned. Leaving the original arch in position is not then a problem. I would suggest that a steel lintel is used when a false arch is required. Catnic produce a solution to both the installation of Soldier and Curved arches and their current brochures are worth reading in this context.

Don't forget that temporary support will be necessary in some cases, and almost certainly where the floor joists are bearing onto the wall in question. Experience and common sense will decide whether or not support is required, but when in doubt provide support or contact the Building Inspector or District Surveyor for advice.

The positioning of the new window has to be carried out in accordance with the current regulations and once again this will have to be approved by the Building Inspector (or District Surveyor, in London). The Regulations dictate the height of the window, the transom height above eye level, the area of glass and the area of openable window required. The Building Regulations themselves are not always the easiest of documents to read and the Building Control Officers at the local council will be pleased to advise you about the requirements. It is best to make an appointment to see them though, as they spend a lot of time out of the office.

Most of the well known manufacturers of standard joinery produce well designed, if not always well constructed, windows. It is as well to check however that sufficient anti-capillary grooves have been cut in the sides and head of the frame to stop the passage of rainwater. The sill should be weathered and throated correctly and

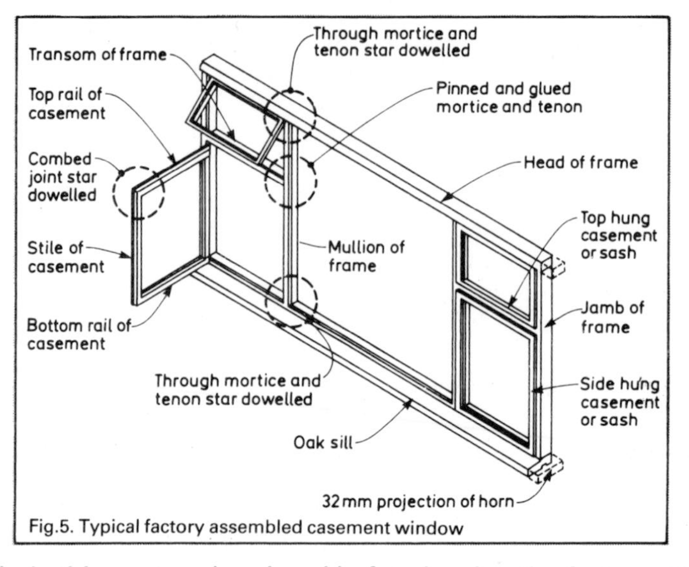

Fig.5. Typical factory assembled casement window

the whole thing properly primed before leaving the factory. Figure 5 shows a typical side hung casement window.

Prior to the London Building Acts at the beginning of the 18th Century, windows tended to be positioned flush with the external elevation of the wall. This, however, helped the spread of flame in a fire, and so the Act specified that the window had to be set back at least 100 mm from the face of the brickwork with a brick reveal. Positioning the window deeper in the opening also helps to protect the timber from the elements and, consequently, rot.

Of course, where the window is made with the sill attached, you have to be sure that the throating of the sill protrudes sufficiently to allow water to be thrown off. In any case, the manufacturer's instructions should be closely adhered to with regards to fixing.

Many older windows did not have a method of damp-proofing to their underside. This tended to cause dampness in the brickwork, a common fault. Before installing the window, a metal or bituminous felt damp proof course should be positioned between the window and the brickwork. If there is a stone sub-sill then this is normally removed as it will almost certainly be of the wrong size once the new window is in place.

The new window may have an oak sill, and although the whole of the window will be primed in the factory, it is unlikely that the manufacturer will have used the proper primer. Aluminium primer ensures that the acids in the oak are separated from the paint film. Be sure to use a good quality primer such as Blackfriars otherwise the paint will not adhere. There should be a high percentage of metal in the primer and it should have a silvery sheen to it when dry.

As in the previous article on window replacement, pointing, plastering, glazing, etc. should all be allowed for in the estimate.

Decorations will once again depend upon the client's requirements, but be sure to properly prepare the window, knot, prime and stop all imperfections prior to decoration.

Allow in the price for making good to any damaged surfaces and for thoroughly cleaning up on completion.

Build-up of estimated cost to replace double hung sash with standard side hung casement window.

As mentioned in a previous Project, it is necessary to consider whether this is a job in isolation, a single opening of which there are several similar but in different locations, or an item of work within a much larger contract.

The prices that follow are for one window in isolation:

		£
a)	Clear and protect area in front garden for scaffold tower (one man ¼ day)	12.56
b)	Supply, erect, maintain, dismantle and clear away 2.00 metre square scaffold tower 6.00 metres high including all necessary scaffold boards; transport on and off site,	
	Scaffold hire, say one week	15.00
	Transport	20.00
	One man one day	50.23
c)	Move furniture and protect as necessary, (one man ¼ day)	12.56
d)	Hack off internal and external reveals to free sash and clear away (one man ¼ day)	12.56
e)	Remove sash from opening and carefully lower to ground for transport off site (two men ⅛ day)	12.56
f)	Remove existing arch and lintel including providing all necessary temporary supports; increase width of opening from 812mm to 2388mm wide by cutting away both reveals; quoin up reveals, cut away for and build in Catnic lintel; construct new arch and leave ready for new window,	
	Four man days	200.92
	Material	88.00
g)	Allow for temporary protection (Provisional)	30.00

h)	Allow for making good pointing around window opening, say	
	2M—one man ¼ day,	12.56
	Material, say	3.30
i)	Supply only new window (446 PWV) including transport,	106.62
j)	Prepare new window prior to fixing including all necessary knotting and priming and removal of horns,	
	One man, ¼ day	12.56
	Materials, say	2.20
k)	Supply and fix frame cramps and cut pockets in reveals for building in,	
	One man ¼ day	12.56
	Materials, say	3.30
l)	Supply and fix damp course between frame and reveal,	
	One man, half hour	3.22
	Material, say	2.20
m)	Offer up window to opening, level up and plumb before building in cramps. bed cramps in cement mortar and replace and re-bed any bricks removed,	
	Two men ⅓ day	33.15
	Material, say	2.20
n)	Point around frame externally in mastic,	
	One man ¼ day	12.56
	Material, say	7.70
o)	Make good all internal plastering and external rendering to match existing,	
	One man ½ day	25.12
	Material, say	8.80
p)	Glaze sashes in appropriate glass with putty and sprigs,	
	Assume 2.5 SM glass	38.00
	One man ⅓ day	16.58
q)	Prepare and paint all exposed timber surfaces to match existing,	
	One man ¾ day	37.67
	Material, say	6.60
r)	Make good all disturbed or damaged decorations internally and externally, *Allow Provisional Sum*	20.00
s)	Clean windows, clear up and remove all rubbish; remove temporary protection, reinstate furniture etc. and leave all clean and tidy to employer's satisfaction,	
	One man ¼ day	12.56
t)	allow Provision Sum for contingencies	50.00
		883.85
u)	Allow additional preliminary costs 12½%	110.48
		994.33

NOTE: The foregoing labour rates have been based on £4.60 per hour plus 40% for profit and overheads.

Project 3
Forming an opening

Many of today's younger families find too irresistible to ignore, the space that forming a through lounge or kitchen/dining room can provide. The traditional design of the 1930s house, with two reception rooms and a small kitchen worked well in its day. However, ideas have changed dramatically during the past fifty years and so too have the needs of the family.

One of the main reasons why this form of home improvement became so popular in the early 1970s was certainly the advent of central heating. No longer did the family have to huddle together in one room to keep warm. With a centrally heated house all rooms became inviting and so if the wall between the lounge and dining room was removed, a feeling of space in one large living room could be achieved.

Many householders are do-it-yourself enthusiasts, but the dangers of anyone with very little or no knowledge of the structural aspects of a house attempting such a job are obvious. Many people, including some builders, take the attitude that all that is needed is a sledge-hammer and an RSJ. They couldn't be more wrong.

Any work to a house which involves a structural alteration requires approval under the Building Regulations or London Building Acts. A drawing should be submitted showing the plan of the ground floor and the position and span of the wall to be removed and beam to be inserted. The upper floors will also have to be shown so that the Building Control Officer and any other interested party can clearly see the layout of the house and, more importantly, the loads that are imposed upon the wall being removed. In addition to this, a section through the opening and a Block Plan of the surrounding houses and streets has to be drawn, all accompanied by the correct fee.

Usually, the application will have to include structural calculations proving that the beam will support the load above.

Now all this sounds very time consuming and expensive, but if permission is not obtained for work of this nature and a competent surveyor makes inquiries when the house is being sold, then this could delay the sale. So it is in everyone's interest to obtain the relevant approvals; the client, the builder and, for safety's sake, everyone who enters the house.

Figure 6 shows the type of sectional drawing that the council would like to see. It is not always possible to know about the foundation details, but the council would require these to be rebuilt around the base of the piers if they were not adequate to distribute the weight above.

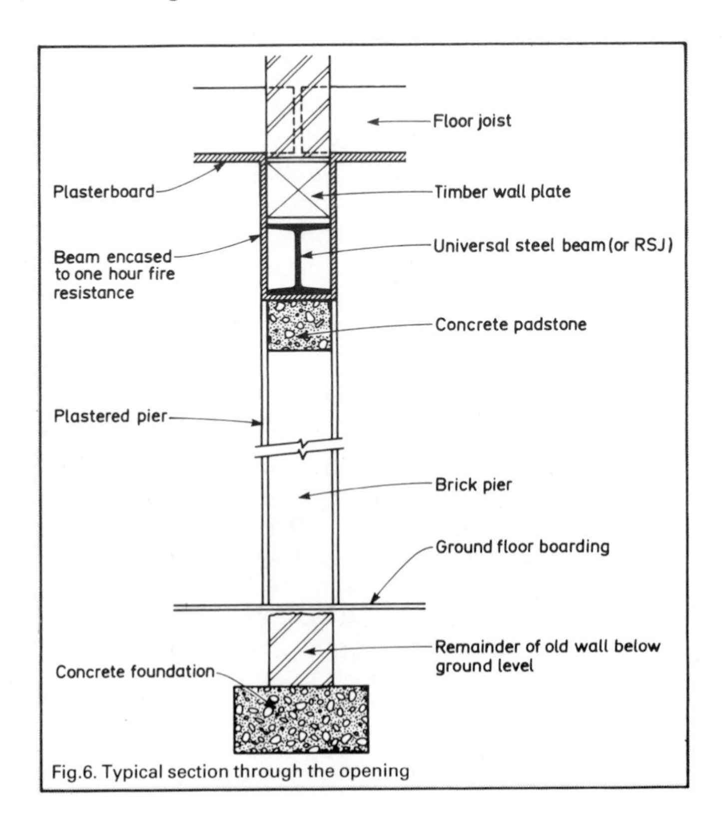

Fig.6. Typical section through the opening

The first stage is to move and protect furniture and belongings in the house. Preparation might even involve the erection of temporary screens over the doorways into the two reception rooms to prevent dust from landing on everything in other parts of the house. Dust, however, seems capable of passing through anything and the occupants of the house should be made to realise that, for a couple of days at least, their home lives will be in disarray.

Lift the floorboards adjacent to where the piers will be and check to make sure that there are adequate foundations below. The Building Inspector may wish to make his first visit at this stage in order to satisfy himself that all is in order. If not then, as mentioned before, it may well be necessary to dig out and construct new concrete foundations around the pier area.

The parts of the wall to be left as piers should be exposed by hacking away the plaster and inspecting the brickwork behind. The wall may be built of blockwork though and its load-bearing qualities will then have to be ascertained by a competent engineer to see if they are capable of carrying the beam.

It is assumed that first floor joists bear onto the wall, and so the floors on both sides of the proposed opening will have to be suported. Scaffold boards laid parallel to the wall (see figure 7) help to spread the point loads on adjustable Acrow-type support props. As will also a scaffold board above the props on the underside of the ceiling.

If the first floor wall above is non-loadbearing, or if the load is taken by the floor joists, then there is no problem when removing the wall. If, however, there is a substantial load still bearing onto the wall then this will have to be supported as well. Figure 7 shows one method of doing this and that is to pin through 300 mm or so above the first floor and support these pins as for the first floor itself. Although construction methods vary from house to house, a timber wall plate will probably have been built in to support the joists. Figure 6 shows such a plate and its relationship with the new beam.

The wall is now ready to be demolished and great care should be taken when this part of the work is carried out. It goes without saying that brickwork is heavy, so start at the top and work downwards. Some builders find it easier to just cut away that part of the wall where the beam and padstones are to be inserted and only demolish the lower part when the beam is ready to carry load.

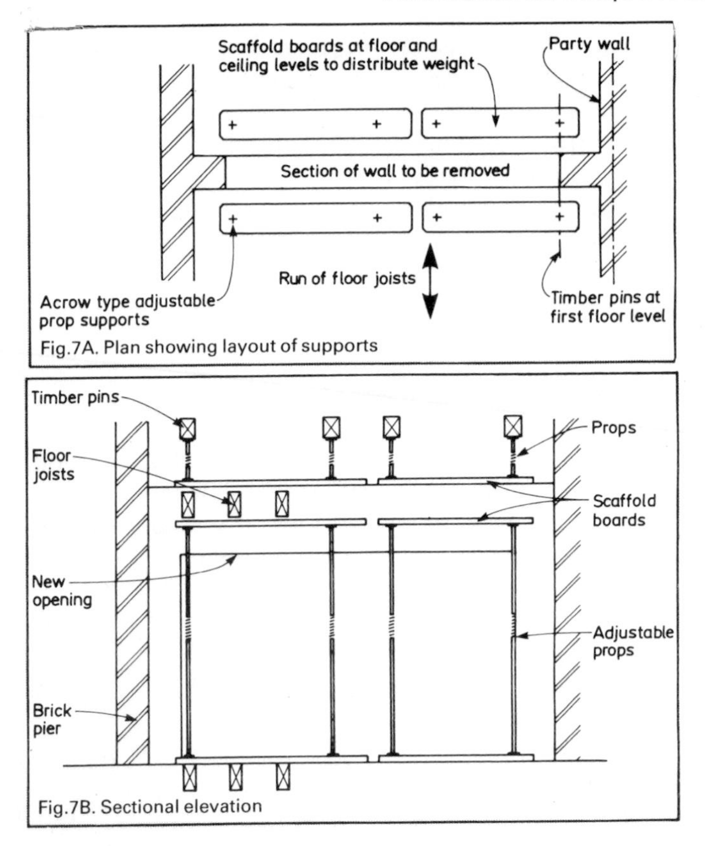

Fig.7A. Plan showing layout of supports

Fig.7B. Sectional elevation

The trouble with this method is that the piers then have to be worked on when they are loaded and the job is complicated by the redundant wall still being in place.

It is recommended, therefore, that the wall is removed and all the brick rubble carted away before reinstatement work starts. The floor can be made good in the opening with a floorboard bedded onto a reasonably dry mix of cement and sand. Be sure that the floor finishes up level.

Sometimes, the beam can span the whole width of the room without the need for piers, but a few points should be noted if this is going to be done.

Firstly, if one end of the beam is being built into the party wall then a Party Wall Award or the written permission of the adjoining owner will be required. This is not necessary with a pier. Secondly, because the span is greater the beam will need to be of a larger section and consequently will be heavier. Lastly, it is almost impossible to plaster the flat wall surface to such a high standard as to totally disguise the position where the wall used to be.

Brick piers in most domestic situations will probably be calculated at 112 mm square with padstones to match and with a depth of 150 mm (two bricks). Proper handling of the beam itself is really a skill that comes with experience. Suffice to say, however, that safety and care are of prime importance.

With the piers properly built, the padstones securely bedded and the beam in place, levels should now be re-checked to ensure that the beam is true. Very minor alterations can be taken up in the thickness of the cement and sand bed between beam and pad. Anything more will have to be adjusted by the insertion of good quality Welsh slate. Slate was formed millions of years ago by compression, so the weight of a steel beam and part of a house certainly isn't going to crush it.

There is likely to be a small gap between the timber wall plate and the beam in some places. Once again, a very dry mixture of cement and sand carefully packed between the two will be adequate unless the gap is large in which case use slate again.

The props should remain in place for about 48 hours to give the whole new structure time to gain strength without the burden of load. When all the supports are removed or during the previous 48 hours, the Building Inspector (or District Surveyor) should be invited to inspect the work. If the beam is encased in plasterboard and skimmed over without his knowledge then he can demand that it is all re-exposed for his inspection.

Use of aluminium edging when the beam has been covered to give one hour's fire resistance will give a neater and truer edge to the finished plasterwork. Don't forget, however, to fix the plasterboard with the correct nails, always galvanised. A highly competent plasterer should be employed as it is very difficult not to be left with a few ragged corners and only an experienced tradesman would know all the ways of disguising such annoying minor defects.

At this stage, cornicework may need to be reinstated by a fibrous plasterer qualified in this type of specialist work.

Remember also that if new plasterwork is to be decorated, only emulsion paint should be used, to give the job a chance to dry out thoroughly and allow for the making good of shrinkage cracks.

Repair and renew the skirting around the base of the piers and prime the new woodwork. Ensure that everything in the house is left clean and tidy and that the client is completely satisfied.

Forming an opening between two rooms

As mentioned in previous Projects, it is necessary to consider whether this is a job in isolation, a single opening of which there are several similar but in different locations, or an item of work within a much larger contract.

		£
a)	Move funiture in both rooms and protect work with dust sheets. Roll back floor covering and protect. Ensure all walkways from scene of work to extrnal area for rubbish disposal are also adequately protected.	
	Labour: one man ⅓ day	16.58
	Plant (dust sheets): say,	10.00
b)	Lift floor boards as necessary to check adequacy of existing foundations including any necessary excavation.	
	Labour: two men ½ day	50.23
c)	Hack off plaster etc as necessary to ascertain wall construction.	
	Labour: one man ⅛ day	6.28
d)	Supply and erect temporary support to upper floor using Acrow or similar telescopic props with suitable timber sole and head plates to spread the load; maintain throughout work and clear on completion.	
	Labour: two men ½ day	50.23
	Plant: say,	21.00
e)	Cut through existing wall (assumed to be half brick wall in flettons in lime mortar) to form new opening; clear rubbish and cart to local tip) save skirting for making good).	
	Labour: three men one day	150.69
	Plant: say,	6.30
	Transport: say,	21.00
NOTE: based on opening size 2.50 M×2.14 M high		
f)	Make good exposed ends of wall or form new piers to correct opening size; prepare seatings for and supply and build in concrete padstones for ends of new beam,	
	Labour: two men ½ day	50.23
	Material: say,	22.00
g)	Make good floor in new opening with timber noggins and new floor boards to match existing,	
	Labour: one man ½ day	25.11
	Material: say,	11.00

h)	Supply, hoist and bed in position suitable RSJ,	
	Labour: two men ½ day	50.23
	Material: say,	50.00
i)	Encase RSJ in expanded metal lathing and plaster or encase in plasteboard fixed to suitable noggins and finished with a setting coat to match existing and plaster piers to match existing,	
	Labour: two men one day; one man ¼ day	113.02
	Material:	39.00
j)	Re-fix and make good skirtings/picture rails etc. to match existing using sound salvaged material,	
	Labour: one man ¼ day	12.56
k)	Prepare for and decorate all new work and make good all damaged existing work,	
	Labour: one man one day	50.23
	Material: say,	11.00
l)	Clear away all rubbish, tools, plants etc. Remove dust sheets, replace floor covering and furniture and leave ready for occupation,	
	Labour: one man ½ day	25.11
		791.80
m)	Allow for preliminaries 12½%	99.98
		890.78

NOTE: The foregoing labour rates have been based on £4.60 per hour plus 40% for profit and overheads.

Project 4
Insulating a domestic roof

During the past 10 years, people the world over have become increasingly aware of insulation and the benefits this can bring to both themselves and the environment as a whole.

It took the world oil shortage and the rapid price increases of all fuels in the middle 1970s to make people, and governments in particular, realise that something had to be done to reduce the country's energy bill.

Double glazing and cavity wall insulation are very popular methods of reducing heat loss through walls, but their cost effectiveness is minimal compared with draught proofing around doors and windows and, of course, loft insulation.

The Government's 1982 Home Insulation Scheme is aimed largely at encouraging the do-it-yourselfer to spend a weekend in his attic with rolls of glass-fibre matting and a torch and the thought in mind that his gas bill will never be as high again. However, not all householders have the time, inclination or sometimes the good health to do the job themselves and so the scheme also provides for the work to be carried out by a contractor.

The amount of grant is relatively small and for most people 66 per cent of the cost, up to a maximum of £69, is all they can claim. Pensioners and disabled people can normally claim more and this is good because these are the sections of the community who are unlikely to be in a position to do it themselves.

To be eligible for a grant, the roof space must be completely uninsulated at the start and the job itself must allow for all pipes, tanks (even hot water cylinders elsewhere in the house) and areas of the roof to be thoroughly insulated.

Too much insulation can cause other problems though, especially if the roof space is poorly ventilated. Condensation is the most important factor and so proper ventilation is vital if a damp, still atmosphere is to be avoided.

Moist, unventilated areas provide perfect conditions for the formation of dry rot. If the underside of roof tiles is not felted or boarded then this problem in minimised, although additional ventilation may still need to be provided. The two cheapest and most popular ways of doing this are to cut holes in the soffit board of the eaves at regular intervals and to insert air bricks into the gable ends at high level (if they exist). Remember to provide insect grilles where holes are cut and behind air bricks. See Figure 8 for details of ventilation. Purpose made grilles can be bought for these situations.

When estimating for a job such as this, it is easy for the contractor to decide on the amount of materials he will need and to add his labour and overheads to arrive at the final price. However, things are never quite as straightforward as they ought to be and it is always a good idea to make a check-list before arriving on site to ensure that everything is allowed for. This is true of all jobs and it will help to reduce errors and omissions in the final price. After all, the contractor not only needs to make a living wage but also a

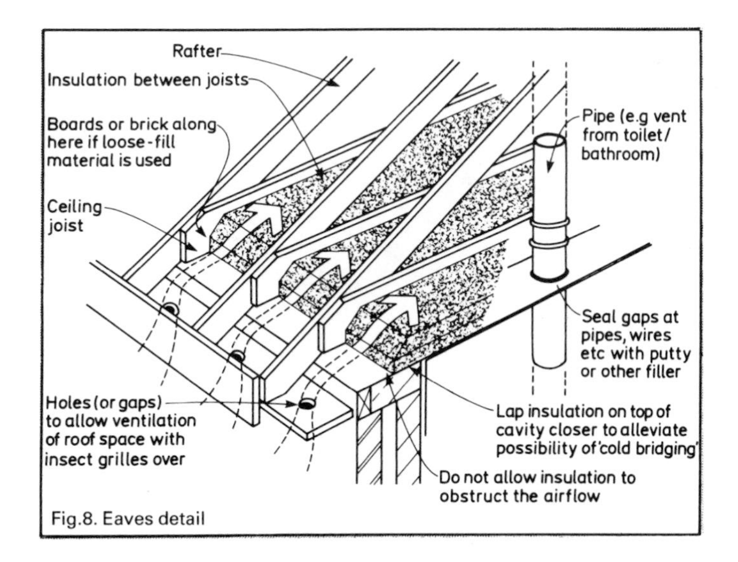

Fig.8. Eaves detail

profit in order to replace plant and machinery and to expand the business.

First on the check-list should be access. A lot of houses don't even have an access door into the roof space and one may well have to be cut open and made good at the end of the job—an expensive item. the grant only allows for this though and not a permanent trap door. If this is required then the householder will have to pay for it himself. There are a number of purpose made loft access doors on the market that simply slot and screw in once the hole is cut. This would probably work out cheaper than making good the hold afterwards.

Is the loft lit by an electric light or will the whole job have to be carried out by torchlight? This will make a lot of difference to the time involved.

The two main types of insulation are glass-fibre quilting on a roll and loose-fill (either mineral wool or vermiculite). Over the past few years, the thickness of insulation required has increased to such an extent that the ceiling joists are sometimes shallower than the insulation itself. When the joists are not visible, walking around in the roof space becomes increasingly difficult. Figure 9 shows the quilting laid across the joists rather than between them and the joists boarded over to allow access to the cold water tanks. If this method is used be sure to fill the gaps at the end of the joists to

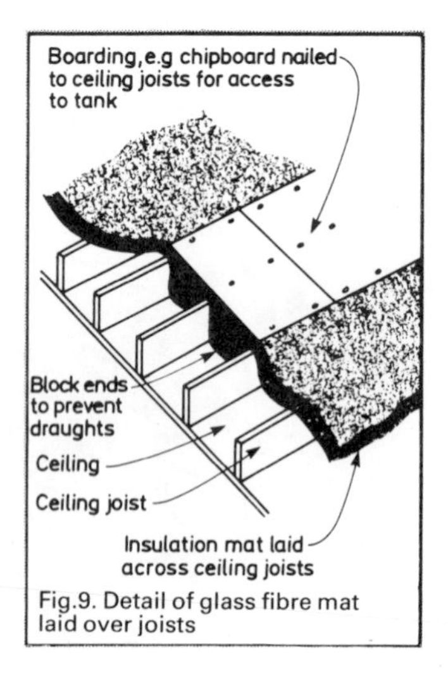

Fig.9. Detail of glass fibre mat laid over joists

stop draughts penetrating between ceiling and insulation. If loose-fill material is to be used in this situation additional timbers will need to be fixed to the tops of joists for either walking on or fixing boards to.

Figure 8 also shows the insulation terminating just short of the eaves. This is to allow the easy passage of air through the new holes in the eaves and so provide adequate ventilation. Also notice in Figure 8 that the insulation stops on top of the cavity wall. If it finished any shorter then "cold bridging" could occur. Without going into too much detail, this causes mould growth to form around the join between ceiling and wall by altering the point within the wall at which moist air condenses on a cold surface and so forms condensation.

Glass fibres have a tendency to get everywhere and can irritate the skin if precautions are not taken. The rolls of quilt should only be opened once they are in the roof space and even then, the access door should preferably be closed. A dust mask should also be worn to prevent fibres from being inhaled.

The loose-fill materials will settle and try to move out of the roof space and out through the holes in the eaves if they are not restrained by boards or bricks as shown in Figure 8.

Take care not to damage the ceiling.

Thicknesses of materials vary according to current legislation. Consult manufacturer's instructions and the Local Authority to determine the amount of insulation required. If insufficient insulation is laid then the client may not receive his grant.

Unless the water tanks are situated a fair distance above the ceiling level it is best not to insulate below them. The heat still escaping from the room below will help to prevent the water from icing-up in cold weather. Insulate also the access door and provide a draught-proofing strip around the seal.

Purpose made insulation jackets can be bought for the cold water storage tank. If the tank is an old galvanised one, give it a thorough inspection for rust and leaks: it may well be getting near the end of its useful life and the householder might appreciate a report on this and an additional estimate for its renewal if necessary. Figure 10 shows one method of insulating a rectangular tank using sheets of expanded polystyrene cut to size. This type of insulation will also act as a removable cover for the top of the tank

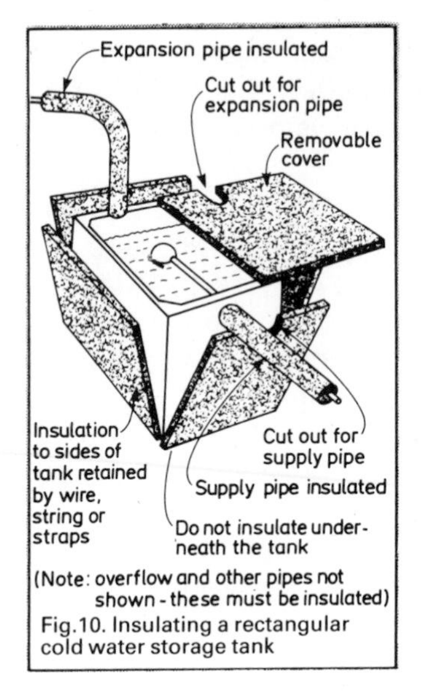

Fig.10. Insulating a rectangular cold water storage tank

to keep out dust and dirt. If the tank is of the circular plastic type, then a special plastic cover should be purchased for this reason.

Pre-formed foam pipe insulation is the best way of dealing with the multitude of pipes found in a centrally heated home. Pipewrap supplied in roll form, which is wrapped around the pipe like a bandage, can also be used. Do not leave any gaps and ensure that the taps and joints are wrapped up well. In this case a central heating expansion tank will also need to be insulated. Pipes that run between the ceiling joists and are below the insulation will need to be separately insulated.

To obtain the grant, the hot water cylinder or tank will also need to be properly insulated. Purpose-made jackets are manufactured in a variety of sizes to suit most modern cylinders. If it is not possible to fit one of these, box-in around the tank and fill with a loose-fill insulation material. Take care not to cover the thermostat and other electrical wiring. Contact the local Electricity Board if in doubt about any of the wiring either in the vicinity of the hot water cylinder or in the roof space itself.

Before leaving the job, ensure that all areas of ceiling, tank and pipe are adequately insulated and that ventilation is efficient and

Glass fibre insulation being laid between joists. The cold water tank is insulated and the pipes covered.

thorough. Don't stand in the roof space with the access trap open and think that ventilation is good. The chances are that a lot of it is coming through the hatch itself.

Insulating a domestic roof space

Note: It is assumed that a trap door exists into the loft space but no loft light is available. Separate adjustments are given at the end to form and make good access and also to provide a new permanent trap door.

As mentioned in previous Projects, it is necessary to consider whether this is a job in isolation, a single roof of which there are several but in different locations, or an item of work within a much larger contract.

		£
1a)	If loft is used for storage this must either be emptied out or time allowed for moving it around the loft space. Allow a *Provisional Sum* subject to site check of two men, half day.	50.23
1b)	Unload, hoist into loft space and lay fibreglass between joists (assume loft size approximately 7.00×7.00 metres) including providing any necessary temporary walkways, including trap door.	
	Labour: two men one day	100.46
	Material: 100 mm fibreglass	166.20
1c)	Supply and fixe standard insulation to C.W.S. tank including lid.	
	Labour: one man ⅕ day	10.05
	Material: say,	14.00

1d)	Supply and fix Celafoam or similar lagging to pipes in loft space (assumed approximately 10.00 metres of pipe).	
	Labour: One man ⅛ day	6.28
	Material: say,	10.00
1e)	Provide ventilation via eaves by cutting openings and fixing mesh insect grilles (assumed twelve provided).	
	Labour: One man half day	25.11
	Material: say,	6.60
1f)	Clear away all surplus materials and leave clean and tidy.	
	Labour: One man ¼ day	12.56
		401.29
1g)	Allow for prelims 12½%	50.16
		451.45

Add—if access to be formed, and make good

2a)	Cut opening through existing ceiling between joists to allow access for labour and materials and clear rubbish from site.	
	Labour: one man ⅛ day	6.28
2b)	Make good opening with all necessary timber noggins, plasterboard and setting coat; make good decorations and leave all clean and tidy.	
	Labour: three men ¼ day	37.67
	Material: say,	11.00
	Transport: say,	5.50
		60.45
2c)	Allow for prelims 12½%	7.56
		68.01

Add—if new trap door required

3a)	Cut opening in existing ceiling including cutting and trimming ceiling joists as necessary; provide trap lining, stops and door and fix in prepared opening. Finish with suitable architraves and paint all new work to match existing and make good any damaged or disturbed adjacent areas. Clear all rubbish.	
	Labour: two men ¾ day	75.35
	Material	44.00
	Plant/transport: say,	11.00
		130.35
3b)	Allow for prelims 12½%	16.29
		146.64

Note: If either Items 2 or 3 apply then the loft will be empty, therefore Item 1(a) will not be required.

Note: The foregoing labour rates have been based on £4.60 per hour plus 40% for profit and overheads.

Project 5
Replacing an external door

Not all doors to fifty year old houses need replacing; the majority are in good condition and have been well maintained and regularly painted over the years. For many, however, the story is far sadder. Poor design, poor construction, poor materials and poor maintenance all help to reduce the life of a door, especially an external door. We start with the external front door because this is probably the most popular one (with the exception maybe of patio doors, which are discussed later) to be renewed.

Everyone likes to be greeted by a nice front door when approaching a house for the first time. It is usually the first part of a house that is seen in any detail and first impressions always mean a lot. The character of a property and indeed of the owners is reflected in it. Is it kept clean and well painted? Is there a brass knocker and letter plate or a chiming bell and aluminium flap? People seem to be spending a lot of their spare cash on new doors even if they have just bought a brand new house, so many of them will spend some extra money on employing a competent builder to hang them.

Magnet and Southerns is probably the first company that springs to most people's minds when replacement doors are mentioned. It produces a reasonably good quality door and its volume of production means that its products are good value for money.

Boulton and Paul is aiming at the same market along with a host of other firms, so competition is fierce.

Hardwood doors are the most popular at the moment, usually with a few squares of glass. These do tend to look out of place in a lot of situations and are sometimes the only piece of stained

French casement frame from Boulton and Paul.

hardwood in an otherwise painted house. Either way, the door that is chosen should be aesthetically well balanced and of a suitable design to allow enough light to enter the hallway. Check that the modern standard-size doors will fit neatly into the existing frame. A few millimetres or so can be planed off the door without affecting its strength or looks. If the door is too small, planting a hardwood lipping along the edge of the door is unwise. Far better to increase the size of the frame (which is likely to be in softwood and so can be painted over on completion).

The frame itself should be checked to see if it needs renewing or repairing.

If the frame does need renewing, the whole nature of the job changes. If standard frames don't fit the opening, the client is left with one of two options. Either the opening has to be altered to suit the frame, or a purpose made frame, and consequently the door has to be made to suit the existing opening.

For alterations to openings see Project 2.

The standard joinery manufacturers also produce a wide range of frames to suit their doors and most of these are a better solution to the problem than trying to hang the new door in the old frame.

A heavy hardwood door will need to be hung on a pair-and-a-half of strong steel or brass hinges and the fixing edge of the frame will also need to be strong enough to support the weight.

Once the door is in place it will need to be treated and lacquered accordingly for it to win its battle against the elements. Draught excluders come in a variety of styles and the client should be made aware of these and given guidance on the one that best suits the situation.

It is better to use the copper 'Atomic' type of draught excluder, as this has a greater lifespan than self-adhesive foam.

Security is of great concern to most householders nowadays, especially as crime (and house burglary in particular) is the only profession currently enjoying a massive increase in student membership! A mortise dead lock or at least a double locking night latch, is a necessity on a front door. These types of locks prevent the door from being opened from the inside without a key. If a break-in occurs elsewhere in the house, the thief cannot calmly walk out of the front door with his ill-gotten gains, he has to try and remove them by the same passage that he entered the house.

Letter plates should also be taken in account, and the hole formed prior to hanging the door. It can be seen, therefore, that there are a number of other items that must be considered when pricing for renewing a front entrance door, apart from the door itself.

As mentioned above, another popular form of door renewal is the installation of patio doors.

Figure 11 shows the typical type of French door arrangement that can be found in a 1930s semi-detached suburban house. Maintenance costs are high and proper redecoration is often neglected. Aluminium sliding patio doors provide plenty of daylight, no draughts, very little maintenance and, if double glazed, reasonable thermal insulation. They can look out of place in many situations and so other forms of door should be considered. Boulton and Paul offers a wide range of external doors. In addition to the sliding aluminium doors it produces good quality sliding timber doors and a range of side hung French-type doors.

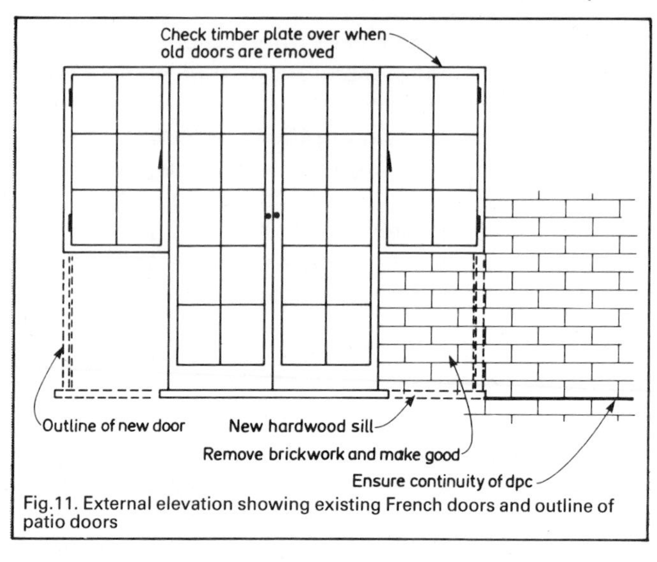

Fig.11. External elevation showing existing French doors and outline of patio doors

Timber is a good natural insulator and condensation, so often a major problem with all metal doors and windows, is kept to a minimum.

As can be seen in Figure 11, if brickwork has to be removed to alter the opening there is no structural problem and only making

Front door in West African mahogany.

The Colonial door from Fourways Homes and Leisure.

good to the reveals is required. The damp-proof course should be altered, if necessary, to run along the underside of the door threshold and care should be taken to ensure that the dpc is not bridged by any external steps or rendering. If the size and position of the doors alter steps are likely to have to be altered too. If it seems that the steps are going to bridge the dpc, leave a gap of 50 mm between them and the wall. Impress upon the client the importance of keeping this void clean of debris and leaves etc.

Check the beam above the old doors supporting the brickwork above. This may be timber and careful investigation will show whether or not it is still structurally sound. The beam may be made of concrete; if so allowance should be made for a timber batten to ensure that curtain track can be easily fixed.

Figure 12 shows sections through typical double-glazed sliding aluminium patio doors including sill, handle and lock. A hardwood

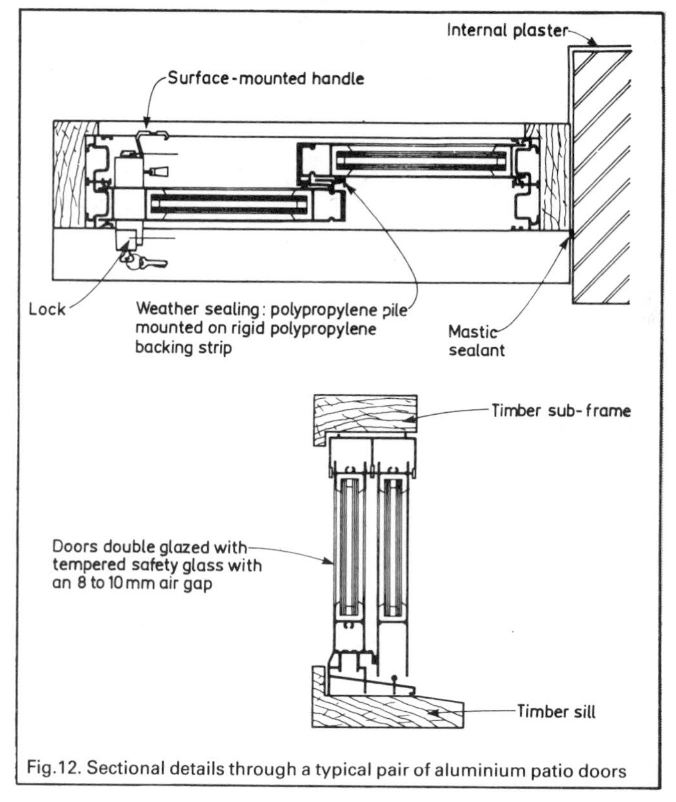

Fig.12. Sectional details through a typical pair of aluminium patio doors

sub-frame and sill are recommended for the fixing of aluminium doors and a mastic sealant should be used between frame and opening.

One problem that many people seem to forget is that, when large patio doors are installed in lieu of old French doors, ventilation may be difficult. Some patio doors are fitted with ventilators but these are not always adequate. This fact should be brought to the house-holder's attention.

Large doors should, of course, always be glazed with a safety glass. Laminated glass is the best solution although this can be extremely expensive. Tempered glass is a reasonable alternative.

Don't forget that the flooring will have to be altered to allow for the new door threshold in place of the original brickwork. Rendering, plastering and decoration will also need to be made good, belongings protected and rubbish cleared away.

Door replacement (as Figure 11)

As mentioned in the previous Projects, it is necessary to consider whether this is a job in isolation, a single door of which there are several but in different locations, or an item of work within a much larger contract.

		£
a)	Move furniture, turn back carpets etc. as necessary and provide dust sheets where required. Protect area externally to door,	
	Labour: one man ¼ day	12.56
	Plant: (dust sheets) say	11.00
b)	Cut back plaster as necessary to free door frame and clear away,	
	Labour: one man ¼ day	12.56
c)	Remove doors from frame and cart away from site,	
	Labour: One man ⅓ day	16.58
	Transport say	6.00
d)	Carefully cut out and remove door frame complete with side lights and cart away from site,	
	Labour: two men ⅓ day	33.15
	Transport say	6.00
e)	Carefully cut down brick panels below side lights and load and cart away all old material,	
	Labour: two men ⅓ day	33.15
	Transport say	6.00
f)	Quoin reveals to match existing including any necessary damp proof course,	
	Labour: two men ½ day	50.23
	Material: say	11.00

NOTE: It is assumed the lintel over is in a satisfactory condition and at the correct height. If this is not the case reference to Project 2 Item (f) will give some guidance on the costs involved.

g)	Supply and fix standard patio pre-glazed door including hardwood sub-frame type HP8 2.382 metres wide×2.057 metres high including all necessary frame cramps etc.	
	Labour: two men ⅔ day	66.81
	Material	730.00
h)	Make good plaster and decorations internally.	
	Labour: one man ½ day	25.11
	Material: say	11.00
i)	Treat new hardwood frame and doors with sealer and point externally with mastic.	
	Labour: ⅓ day	16.58
	Material: say	11.00
j)	Make good flooring to new opening width internally including all necessary noggins etc.	
	Labour: one man ¼ day	12.56
	Material: say	7.00
k)	Clear away all rubbish, replace furniture and carpets and leave clean and tidy.	
	Labour: one man ¼ day	12.56
		1090.85
l)	Allow for prelims 12½%	136.36
		1227.21

If the property is old it is wise to allow a further sum of say £150.00 to cover unforeseen problems.

The foregoing labour rates have been based on £4.60 per hour plus 40% for profit and overheads.

Project 6
Replacing a soil pipe

There are three main ways of carrying soil and waste from the house into the drainage system below ground. These are known as the two pipe system, the one pipe system and the single stack system.

The two pipe system is used in many older houses and basically involves the soil stack taking the discharge from the WC or WCs and a separate downpipe taking the waste from baths and basins. This system ensures that sewer gases cannot enter the house, by the provision of 'breaks' in the waste pipe at the hopper head and gully, and ventilation of the soil stack.

The one pipe system was developed between the wars and generally used in multi-storey buildings. This system relies on the traps (usually of 76 mm) to prevent gases from entering the house. This method is more expensive if used in two storey buildings, as the majority of the traps in such a system need to be ventilated to reduce the risk of siphonage.

The single stack system is a true one pipe system as the appliances are ventilated only by the stack. This method is really only suitable for buildings of no more than two storeys—the average family home.

All three are still installed today in various situations, but this article only discusses one of them in detail: the single stack method, which is probably the most commonly used in single unit domestic dwellings.

The current *Code of Practice* covers new and replaced installations to all sizes of domestic building and should be readily available to both the designer and builder.

Unplasticised pvc is the most popular material for drainage above ground and is manufactured by many companies such as Terrain, Osma, Marley, etc. If the water to be discharged exceeds 60 deg C (as in the case of some washing machines) then polythene or polypropylene should be used. Most manufacturers produce comprehensive literature about the choosing and installing of drainage and waste pipes and it is well worth consulting such brochures before starting the job.

One of the most important points to note is that most householders will not be able to do without their baths, basins and toilets for any length of time and, unless they are on holiday when the work is being done, either temporary wastes will have to be provided or the whole job will have to be completed within one day. Getting the job done in one day is only really feasible if no major alterations are taking place to the sanitary system as a whole. In this case most of the work is likely to be taking place outside the house and so many of the problems associated with working in occupied premises do not exist.

The old pipe will probably be 102 mm diameter cast iron and although the actual removal of this pipe is unlikely to present any problems, the joints from various wastes and WCs probably will. It is recommended that a tower scaffold is used for gaining access to all sections of the pipe. The erection of any form of scaffold may prove a problem however, and although working from ladders may be the only way, it is not a particularly good way of tackling such a job. The savings of not using a proper working platform may be out-weighed by the extra working time needed to carry out the job on ladders.

It may be possible to run the pipe internally. The Building Regulations state that the pipe should only be run externally in buildings of no more than three storeys high. (This regulation, however, only relates to new buildings).

A single stack system can only be used if it conforms to certain requirements however, and if either of the other systems is to be replaced with the single stack, or if additional appliances are being run to the stack, these rules have to be adhered to.

A brief summary of the main recommendations is as follows:

1. The bend at the foot of the stack has to have a radius of at least 150 mm. Therefore, if this is not the case then renewal of part of the drainage pipe below ground will also be necessary.

2. Vertical distances between the lowest branch connection and the invert of the drain should be at least 460 mm for a two storey house. If this is not so, then the ground floor appliances have to be connected directly to the drain.
3. 'P' traps should be used on basin wastes and these should be 32 mm diameter and be no longer than 3 m to prevent self-siphonage.
4. Bath wastes should be 38 mm diameter and can be run through either 'P' or 'S' traps. Self-siphonage is reduced owing to the flat bottom of the bath being able to refill the trap.

The current *Code of Practice* sets out the recommended positions of entry into the stack along with all the other relevant points and it is as well to follow this carefully before and during the job.

Figure 13 shows a typical section through the external wall of a two storey house and an internal soil stack. Although, as mentioned, it is not always necessary to erect the pipe internally if the existing pipe is external, there are a few advantages in doing this. External pipework looks ugly and also is prone to frost attack if water is allowed to lie in any part of the pipes.

The Buildings Regulations and Code of Practice dictate that the stack should be terminated at a height of not less than 900 mm above the top of any window within 3 m from the vent pipe horizontally.

When the pipe is reconnected to the wastes, the joints should all be properly made and the Code of Practice advises on the different methods for various pipe materials. All traps and pipes should be thoroughly checked and renewed if necessary. Remember to provide rodding eyes for access at all changes in direction. It is also advantageous to provide easily removable traps.

Fig 14A shows connection of pvc to a metal pipe of similar diameter. Connecting pvc to a glazed vitreous clay drain can be achieved by solvent welding a caulking ring to the base of the new stack and joining to the underground clay drain using a 1:3 cement/sand mix. An improved key can be obtained by applying solvent weld cement to the base of the pvc pipe and coating it with a thin sprinkling of sand (see Figure 14).

Figure 16 depicts the various stages of connecting a new pvc waste pipe to the new stack.

In some older houses the walls were built in such a way as to try and disguise the presence of downpipes as shown in Figure 15.

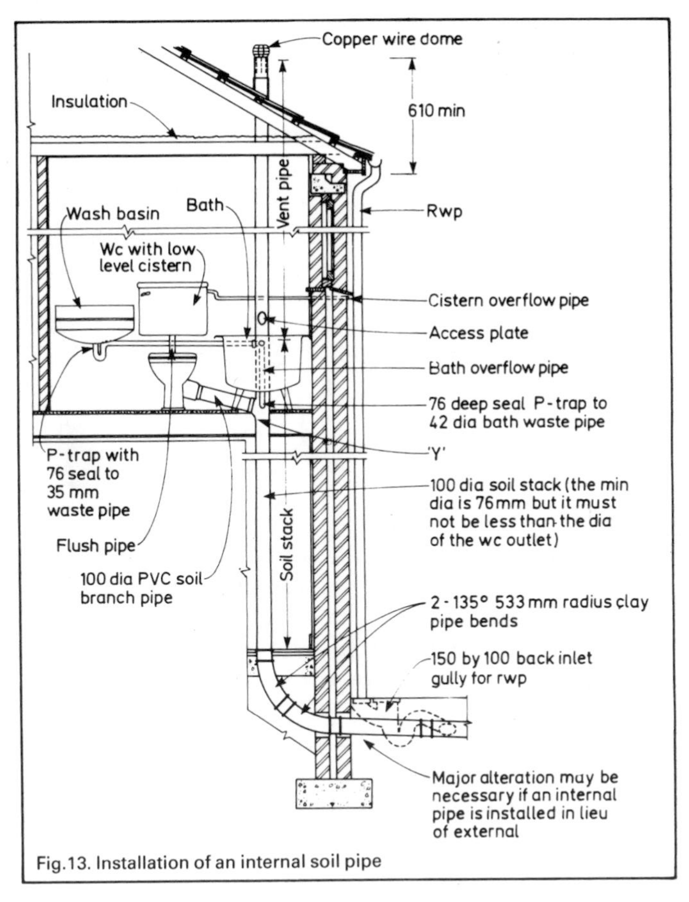

Fig.13. Installation of an internal soil pipe

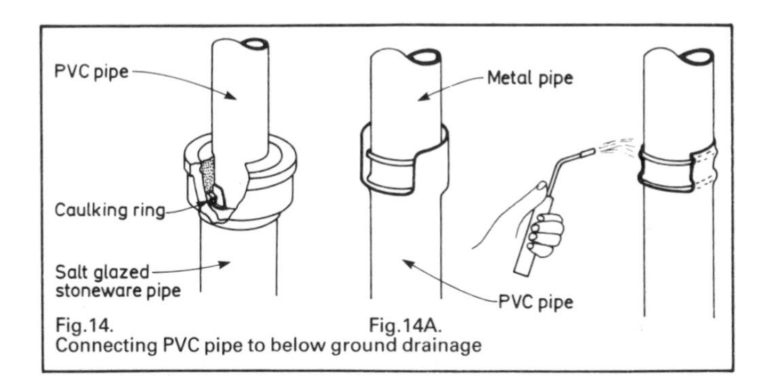

Fig.14. Fig.14A.
Connecting PVC pipe to below ground drainage

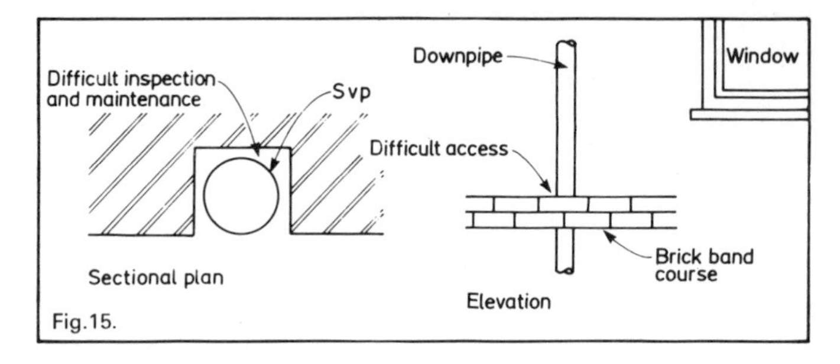

Fig.15.

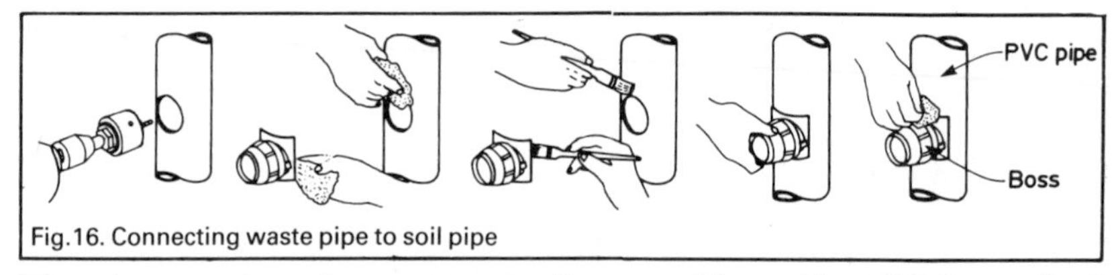

Fig.16. Connecting waste pipe to soil pipe

The pipe need not be automatically repositioned but if it is cracked at the back, waste water may well have penetrated the brickwork. As the construction of timber floors above the ground level is often unventilated there is a distinct danger of an outbreak of dry rot occurring in floors, skirtings, etc. if the pipe has been leaking for any length of time.

When fixing, be sure that any sections of pipe requiring painting or other protective coatings (cast iron, etc.) have a free space for access all round them. This also applies if the wall itself has to be painted or maintained. If the pipe is likely to be badly neglected, as may be in the case of Figure 15, redesign may be necessary.

Many pipes, especially in horizontal situations, are not provided with adequate support. If manufacturer's instructions are followed, however, this will not be the case of the new pipework. Fixings may be in different locations than before and making good the brickwork or rendering should be allowed for in the estimate.

As mentioned, drainage above and below ground is subject to various regulations and, if any changes are being made, will need to be passed by the relevant authorities. Depending upon which part of the country the job is in, the local authority Building Inspector or Environmental Health Officer will make an inspection and will usually require the work to be tested. The current *Code of Practice* lists all the relevant tests.

When the new stack is in place and has been tested and passed, other areas will need to be made good. The path or patio adjacent or a rendered plinth at low level may have been damaged during the course of work. If new wastes have been installed then the wall itself will require making good along with internal plastering. Protect floor coverings, basins and baths, etc. from damage and check on completion to ensure that no damage has occurred and that internal wastes do not leak.

If, after you leave the job, the client complains of poor clearing up or leaking wastes then you are unlikely to be recommended. Consideration, courtesy and reliability are as important to many clients as good workmanship and competitive pricing.

External Soil Pipe

The following price is for one stack in isolation and will necessariy require adjustment should the number off increase or the work be part of a larger contract.

		£
a)	If work is to be carried out from a tower scaffold allow for erection, dismantle and hire, say (Provisional Sum)	40.00
b)	Disconnect WWP flush pipe and remove w.c. pan, Labour: ¼ day	12.56
c)	Disconnect bath and basin waste including removing bath panel as necessary. Release pipes through external walls by cutting away as necessary, Labour: one man half day	25.11
d)	Free joints, dismantle, take down and clear away existing cast iron soil stack, Labour: two men ⅓ day	33.15
e)	Supply and fix new 100 mm uPVC stack complete with branch, w.c. bend, w.c. adaptor, waste connections, plastic vent dome, traps and joint to existing drain, Labour: two men ½ day	50.23
	Material:	98.00
f)	Make good all work disturbed and replace bath panel etc, One man ¼ day	12.56
g)	Clear up rubbish and cart away. Labour: one man ¼ day	12.56
	Transport: say	11.00
		295.17
h)	Allow for prelims 12½%	36.90
		332.07
	If no scaffold use	**287.07**

If the stack is internal considerable additional work is involved such as releasing vent from roof covering and probably replacing the lead or similar slate, removing, repairing and replacing pipe ducts with possible making good of plastering and decorations, taking up flooring and replacing, making good ceilings after removal and replacement and gaining access to the bottom of the stack prior to entry into the drainage system. Further difficulties could arise during installation due to access problems requiring short lengths of pipe and consequently more fittings and joints.
It is not possible to evaluate the foregoing with any degree of accuracy without a particular job in mind, however, assuming a situation similar to Fig. 13 the following should be taken into account.
Assuming bottom of stack enters a stoneware pipe at G/F level:
It may be necessary to make alternative arrangements for a period of 48 hours in respect of toilet facilities; no allowance has been made in the following prices for this eventuality.

a)	Additional labour in removing existing stack, Labour: two men ¾ day	75.35
b)	Additional labour in fixing new stack, Labour: two men ⅓ day	33.15
c)	Additional material (slate in lieu of bend, straight couplers, say 2)	20.00
d)	Replace duct panels etc. make good floors, ceilings, plaster and decorations etc, as necessary, Labour: one man one day	50.23
	Material: say	17.00
		195.73
e)	Allow for prelims 12½%	24.47
	Additional Cost	**220.20**

It should be noted that the overall time scale will be affected by the type of labour and number of operatives employed, particularly in respect of the final making good and possible re-decoration which by it's very nature cannot all be done on the same day. Due allowance for this should be made to cover additional visits, lost time etc.

The prices are based on £4.60 per hour for an eight hour day plus 40% for profit and overheads.

Project 7
Converting a loft space

It is assumed that the builder will have already discussed with his client the reasons and feasibility of installing a room or rooms in the roof of a building. But, however, it is best not to take the client's instructions as gospel, as there may well be a more suitable and economical way of dealing with his space problems.

Some householders seem to take it for granted that the only way of improving or expanding their home to suit their needs is by extending either upwards or outwards.

As a builder you may also be expected to be an architect, quantity surveyor, building surveyor, etc., all rolled into one. It is important to listen to your client's reasons for the proposed loft extension and you may be able to recommend that he employ a professional person in some capacity to help solve his problems. Where planning permission and Building Regulations approval are concerned it is best to consult someone who deals with these situations as an everyday occurrence.

Firstly, you should ask yourself if it is a feasible proposition to convert the existing loft space into habitable rooms. There will be problems with heights of rooms and existing roof layout.

Planning permission may not be required if the dwelling has not been extended at all over the last 30 or 40 years. This would be known as a permitted development and most planning authorities would not be interested in the extra room that it creates as long as this is not too extensive. Most local authorities would be pleased to offer further advice. In order to be classed as a habitable room the conversion would need to have windows. Dormer windows are the usual answer as they not only give the required amount of light and ventilation but also tend to increase the headroom over some of the area. Dormer windows will be dealt with in more detail in Project 9.

Dormer windows will, of course, alter the external appearance of a building and in this respect planning permission is usually required, especially if this is performed on the front elevation. Many authorities consider this too to be permitted development and while they would take the official line that planning permission should be obtained they do not normally enforce this. Experience has shown however that it is wise to send the local planning authority copies of the drawing anyway and ask them to reply in writing that they do not require a formal planning application. This will not attract the usual fee but will give the client peace of mind and will alleviate the possibility of any problems arising if and when the property is sold.

The other main consideration for enlisting professional help is that of Building Regulations (or approval under the London Building Acts in the London area).There are many regulations regarding the conversion of a roof space into habitable rooms and probably the most important and potentially the most expensive, is that of the fire regulations. These regulations will be dealt with in more detail in the second Project in the loft conversion series. The local building control officer or district surveyor will almost certainly require calculations for the various structural alterations that will undoubtedly occur.

The cutting of purlins for a dormer window and the trimming of the ceiling joists for the staircase will once again be dealt with in the later articles. Most purlins are supported by struts running diagonally to a central spine partition or some other load-bearing part of the structure. In order to convert the room these struts will need to be moved or removed and the purlins and rafters will need to be supported in some alternative manner. It is usual to enlarge the sectional size of the strut and to run it vertically between the purlin and the floor, with the floor suitably strengthened. Structural calculations will of course be needed to prove the suitability of the construction. The existing ceiling joists may also need to be strengthened or replaced and this too will require calculations.

Figure 17 shows a section through the roof space and the various elements of construction.

When it comes to the actual building work itself, the window is

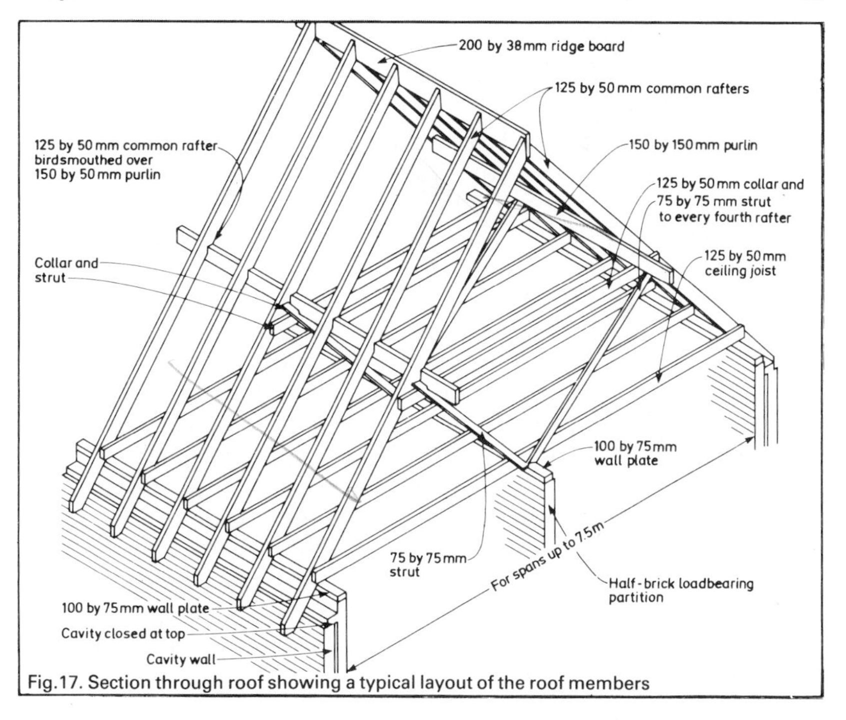

Fig.17. Section through roof showing a typical layout of the roof members

normally the first thing to be installed. This part of the work will require a good system of scaffolding and this may need to be taken right over the existing roof with a temporary weatherproof covering. The Construction Regulations should be complied with regarding the scaffolding and the whole thing erected so as to cause minimum inconvenience to the householders.

Magnet and Southerns B236T dormer-window.

Sometimes the existing roof tiles can be reused on the cheeks and slopes of the dormer window but this once again is dealt with in Project 9.

Standardisation of the building materials i.e. using 100 mm by 50 mm treated softwood members throughout and dimensional co-ordination of the spacing of such timbers is very important to achieve a trouble-free and quickly-assembled timber structure. A distance of 600 mm between the studs will allow for 1200 mm width of plasterboard. Use of the same size of timber throughout, while initially slightly more expensive, does mean that there will be no "ragged corners" when the job is complete.

An important factor to remember is that if the size of the ceiling joists has to be increased to allow for the additional loadings that will be imposed on the new floor, the ceiling height of the new room may be reduced. The regulations state that two-thirds of the floor area should have a ceiling height above it of at least 2.3 m. This means that the other one-third can have low or sloping ceilings. The flooring, or rather the ceiling, joists that are in the roof void parts of the loft would not need to be increased in size as there will be no additional superimposed loadings in these areas.

Hardwood traditional spindles and capped newel from Magnet and Southerns.

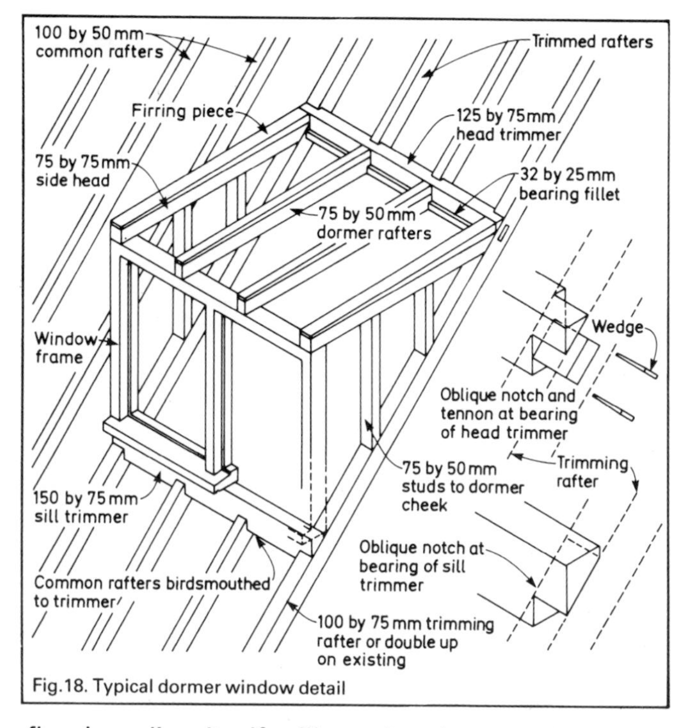

Fig.18. Typical dormer window detail

The floorboarding itself will need to be tongued and grooved chipboard or similar in order to comply with the fire regulations. Tonguing and grooving or covering existing floorboards with hardboards prevents the passage of smoke in the event of a fire.

Insulation for a converted roof space is obviously far more difficult than insulating a conventional roof void. The areas that still remain in a void will need to be insulated in the usual manner and 100 mm glass fibre quilt can be laid between the joists. The newly formed stud partition walls and sloping areas of ceiling are best insulated with glass fibre batts as these can be held in place more easily while the work proceeds.

One of the problems with insulation is that warm air does tend to rise, so that a newly converted loft room can become very hot and stuffy. A good form of ventilation is therefore very important. Building Regulations state that the openable area of window has to be at least 5 per cent of the floor area. It is recommended that this amount be increased, especially where a half-hour fire-resisting self-closing door is not required between the new room and the new staircase. If there is no such fire-resisting enclosure then the warm air will fill the attic rooms even more so.

As can be seen there is a wide variety of aspects to be considered when planning a loft conversion.

Loft conversion

As the descriptive matter for this subject has been spread over three Projects we will also deal with the costing on a similar basis.
This Project deals with the internal loft alterations and the following with staircase/access, dormer window and internal/external finishings. The prices quoted are for a single loft conversion.
We have based this series on house type Figure 17 and assumed a single room only approximately 7.00 x 5.50 metres with one dormer window.

		£
a)	allow to clear out loft space,	
	one man: half day	25.11
b)	Re-site coldwater storage tank and heating header tank,	
	Two men: 1.25 days	125.58
	Material: (pipe, lagging, fittings)	60.00
c)	Additional ceiling joists (125 x 50mm - 15 No.)	
	Two men: .75 day	75.35
d)	Plate to support purlin (2 No.)	
	One man: .33 day	16.58
	Material:	27.00
e)	Remove existing collars and struts and adapt and re-fix in new positions,	
	Two men: one day	100.46
	Material:	33.00
f)	Supply and fix studding for walls including noggins and framing for cupboard doors etc.	
	Two men: one day	100.46
	materials:	180.00
g)	Supply and fix ceiling joists to loft room,	
	Two men: .25 day	12.56
	Material:	53.00
h)	Supply and fix tongued and grooved chipboard flooring,	
	Two men: one day	100.46
	Material:	230.00
i)	Supply and fix plasterboard and insulation to walls and ceiling for direct decoration,	
	Two men: 1.25 days	125.58
	Material:	370.00
j)	Supply and fix sundry vent trunking, grilles etc.	
	One man: .50 day	25.11
	Material: say	33.00

k)	Allow provisional sum for extending electrical services into loft to provide three light points and four twin socket outlets,	200.00
l)	Allow provisional sum for two fan assisted electric convector radiators including necessary wiring,	150.00
		2243.25
m)	Allow for preliminary abd 12½%	280.40
		2523.65

Project 8
Loft conversion II: staircases

The staircase is usually the last part of a loft conversion job to be done, especially in occupied premises, because dust and dirt can be prevented from entering the main building if this is left until the end.

The position of the staircase is largely governed by the existing landing at first floor level and also by the slope of the roof above. Headroom can quite easily be restricted by the slope of the rafters and it may be necessary to have a complicated staircase with half or quarter landings and maybe even winders.

Some local authorities demand that the new attic rooms have a fire-resisting enclosure leading directly to the outside air if the room is at second storey level or above. This means that the habitable rooms at ground and first floor level must be protected from the stairways by half hour fire-resisting enclosures and the doors must be half hour fire-resisting and self-closing. The ground and first floor ceilings are also required to be half hour fire-resisting in this instance.

Some of these regulations can sometimes be waived by the local fire authority or building control department and a modified half hour fire-resistance can be obtained. It is no good having a half hour fire-resisting enclosure if people prop the doors open and remove the self-closers completely. Not many people like the idea of having 25 mm door stops to their living room and bedroom doors and the ugliness, weight and expense of half hour fire-resting doors are prohibitive; a modified half hour fire-resistance is sometimes acceptable. A modified half hour fire-resistance means that only a self-closer is required on the door. This will normally give twenty minutes of protection which can usually be considered adequate for the occupants of the house to escape in the event of fire.

To upgrade a floor to be half hour fire-resisting will normally require hardboard to be fitted over a normal softwood boarded floor to prevent the passage of smoke. This too can be very expensive and can involve the householder in a lot of upheaval.

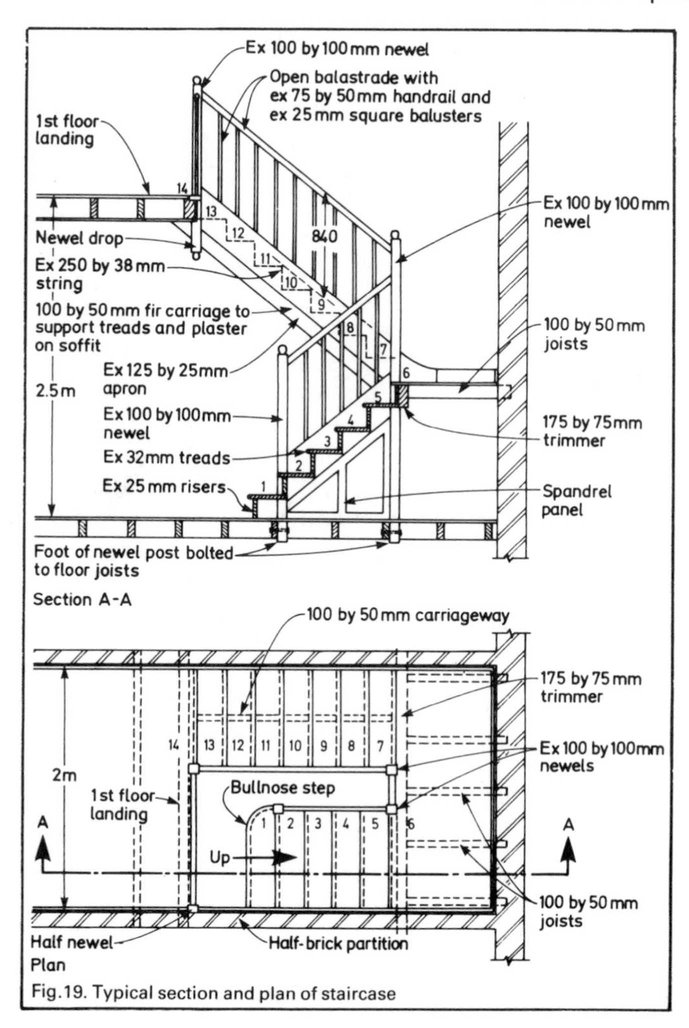

Fig. 19. Typical section and plan of staircase

Once again this regulation can sometimes be waived. It is important to let the client know that these regulations do exist and what might at first appear to be a very feasible and quite straightforward job could well turn out to be a costly and difficult one. The only way of compartmentalising the attic room from the staircase is sometimes to provide a fire-resisting enclosure in the room itself. This once again can look very ugly and can completely alter the effect that might originally have been desired. It is wise to consult with the local building control officer early on in the planning stage of the job to see what his thoughts might be.

In all existing properties it is very likely that a purpose made staircase will need to be manufactured. There are many component joinery workshops around and most staircases can be made to order for a reasonable price. When ordering a staircase you will also need to consider handrails, bannisters, balustrades as well as skirtings and various finishing beadings.

Figure 19 shows a typical plan and section of a new staircase in relation to the rest of the house.

Calculations will need to be prepared by an engineer to show that the floor above can be trimmed to form the right size opening for the staircase. As long as the strings of the staircase can be adequately secured and the newel post supported it is normally only necessary to double up the joists around the first floor ceiling. If the staircase does not comply with all the regulations as far as headroom is concerned the local authority does have the power to waive these regulations if it means that to conform with them would be an impossibility.

The ground floor may also have to be strengthened to support the foot of the staircase or the newel post and allowances for this should be made in estimate.

It is sometimes difficult to estimate accurately without causing too much initial damage to the property and if this is the case the client should be made well aware that some minor extras may be incurred. It is best not to be too vague about the price of the job but do be sure that the client is informed if you cannot estimate for something. Most people appreciate that an accurate price on all matters is not always possible.

If one of the rooms in the newly converted loft space is to be used as a bathroom, as is often the case, plumbing and waste, etc., will also have to be considered. Probably the most important thing is to ensure that there is sufficient head of water to give adequate pressure for all the appliances and fittings. This will probably mean moving the cold water storage tank to a higher level inside the roof void and a great deal of copper piping and preformed insulation will be required.

The client's electrical requirements will also need to be known. How many socket outlets does he require? Do they need to be single or double? You should advise him that they are all switched and the type of fittings both for the power and lighting will

A different approach to loft access, using a bought-in staircase.

probably need to match those in the remainder of the house. Allow for all the clips, brackets, etc., that are inevitably required for all electrical and plumbing installations.

If the house is centrally heated this installation will need to be extended, with maybe a heated towel rail in the bathroom. It is easy to think that a few more radiators can be added on, but is the boiler capable of coping with this extra load? If the client has to buy a new boiler this will alter dramatically the cost of the job. Many new boilers are costing as much as £800 at today's prices and it may well be cheaper and easier to consider a different form of heating in the new rooms.

Extra support may be needed.

Another factor to be considered when converting a loft area is that lots of extra storage can be gained by using the voids behind the new studwork walls that will be no good for floor area. Rather than have just an access panel into these areas from the room itself they can be properly boarded out, preferably with blockboard and a nice set of doors and architraving used to make them into proper cupboard areas. Extras such as this will obviously add to the cost but the additional storage space will be worth it in the long run.

Storage can also be used underneath the new staircase if this is practical and the possibility of built-in wardrobes in the bedroom may also prove to be worthwhile.

Generally speaking, things should match the existing as much as possible and all areas of timber that will not be seen (especially structural) should be properly treated prior to their installation. Where the pricing of such a job is to be competitive it is important that the client gives some form of specification to work to.

There are many builders around who will cut corners at every given opportunity and the client should be made aware that unless he specifies, or employs a professional person to specify, the exact extent of the work and materials to be used prices will differ quite considerably.

Loft conversion - 2

In the last Project we dealt broadly with the general conversion of the loft space into a room. This time we deal with access into the room followed in Project 9 with the dormer window and all finishings.

		£
a)	Form opening in ceiling including providing and fixing trimmer joists, apron linings etc. and making good existing ceilings,	
	Two men: one day	100.46
	Materials:	28.00
b)	Supply and fix purpose made dogleg staircase in softwood,	
	Two men: one day	100.46
	Materials[	810.00
c)	Construct half landing on 100 x 50mm joists covered with 25mm tongued and grooved flooring,	
	Two men: half day	50.23
	Material:	33.00

d)	Provide and fix open balustrade including newels and return at loft level,	
	Two men: .75 day	75.35
	Material:	205.00
e)	Construct spandril framing and cover in plywood including provision of small door and frame for access to understair storage area,	
	Two men: .75 day	75.35
	Materials:	88.00
f)	Sundry items including forming bearings in wall for half landing, making good existing finishings, take up existing flooring to fix newel base, additional timbers in floor to carry new staircase,	
	Two men: one day	100.46
	Material:	44.00
g)	Softwood carriage to new stair,	
	One man: .25 day	12.56
	Material:	13.00
h)	Line underside of staircase with plasterboard,	
	One man: .25	12.56
	Material:	16.00
		1764.43
i)	Allow for prelims 12½%	220.55
		1984.98

Project 9
Loft conversion III: Planning a dormer installation

The dormer window is the only part of a loft conversion that will actually be seen from the outside of the building. Therefore it is very important that it should look aesthetically correct as well as performing the function for which it was intended, in the most practical way.

The Building Regulations state that an area equivalent to one tenth of a habitable room should be window area. Half of this window area should be openable. Therefore the size of the window is governed by the size of the room and vice versa. The window should match those of the existing house even if it has to be purpose made.

To produce a good-looking dormer window it is best to try and achieve some sort of pitch to the roof instead of the more common felted flat roof. The drawings show three different methods of achieving a pitched roof on the dormer that would all add to, rather than detract from, the overall appearance of the property. It is possible that all three could create different ceiling heights in the room itself so this too should be taken into consideration.

When planning the installation of a dormer window remember that although accessibility during construction will be relatively easy because of the scaffold, when this has gone accessibility for maintenance may not be quite so straightforward. If there is no alternative but to use a flat roof then a metal sheet roof will outlast a felted roof by many years. Code 4 lead or copper are probably the best types of finish for these circumstances. They both require little maintenance and can normally be left uninspected for many years.

When the pitches of the dormer, or even the cheeks, are tiled it is preferable to try and rescue some of the tiles from another less visible slope of the building. If the roof coverings are in a poor condition anyway they can all be renewed at the same time as the installation of the dormer. If the dormer is on the front of the building the preferred course of action is to strip the rear roof

coverings and reuse these tiles or slates around the new dormer. The rear slopes can then be reclad in the usual way.

The apex of the dormer should not rise above the apex of the existing roof. If the correct internal ceiling heights cannot be achieved in any other manner a rethink on the whole possibility of a loft conversion is advisable. Local planning authorities will almost certainly require a formal planning application and listen to the views of neighbours if the overall height of the property is to be increased.

Although building control officers and district surveyors will rarely leave their offices to visit the job if they have not received a formal application, they are normally only too willing to give help and advice if an appointment is made with them on their home ground.

There are other regulations regarding the positioning of a dormer window in relation to the party walls, especially on terraced and semi-detached houses. Sometimes the party wall has to be rebuilt to form a parapet; this will not only increase the cost of the job but also entail the long and expensive task of obtaining a party wall award. The height of any adjacent chimneys may have to be raised to comply with the various Building Regulations or London Building Acts.

The party wall itself inside the roof space may also require upgrading to meet regulations. Some roof spaces in older houses do not have walls between properties since most of these were not provided with access into the roof space from the house itself.

The soil and vent pipe from the sanitary appliances normally rises above the eaves of the building and the vent terminates a few feet above eaves level, with a wire or plastic balloon on top. If the dormer is to be built on this elevation then the pipe will probably have to be either extended in height or rerouted completely. This is another item that is frequently overlooked when pricing for this kind of work but it can prove to be an expensive mistake if the builder does not allow for it.

Most dormer windows will need to have some means of disposing of the rainwater that will fall or collect on them. The guttering should match that of the existing house and a downpipe, either connecting to the main rainwater system or discharging over a gutter, will need to be provided. A poorly planned system of rainwater disposal can spoil the whole effect of a nice looking dormer window so this should be carefully thought about prior to beginning the job.

A dormer window will inevitably weigh quite a bit more than the existing area that it is to replace. Many areas will need to be strengthened and the local authority is bound to require structural engineer's calculations before approving the Building Regulations application.

The purlin will need to have a section cut out of it and will therefore need to be resupported by properly sized struts, etc. The opening for the dormer will need to be correctly trimmed and the rafters on either side will almost certainly need to be doubled up. The additional weights on the floor below will mean that this too

Dormer window detail on a Scandia-HUS timber frame house.

will have to be strengthened. Sometimes the insertion of a steel beam is the only way to provide such strength.

Calculations for this should be prepared by a structural engineer; if the builder tries to price for this work without knowledge of the results of the calculations his estimate will almost certainly be very inaccurate.

For the external parts of a dormer window, it is always wise to use lead or zinc lined valleys and flashings where possible, or tiled valleys if these match the existing building. It is good policy on the part of the builder to inspect the remainder of the roof coverings when the work is in progress and to advise the owner of any repairs that may be needed. Most houses will need a few minor roof repairs carried out and while the scaffold is in place and roofers are on site it is a good idea to leave the roof in tip top condition.

For dormer windows in more exposed locations it is wise for a householder to invest in double glazing while the work is in progress; this will be more economical than secondary glazing at a later date.

Internal joinery such as doors, skirtings and architraves should match those in the remainder of the house and allowance should be made for priming all new woodwork or completely decorating it in accordance with the client's wishes. It is not good practice to wallpaper newly plastered areas unless a de-humidifier has been in the room for a few days. A couple of coats of emulsion paint should suffice for six months or so and will allow for any shrinkage, cracks, etc., to be easily spotted and made good. Remember that your client will notice any minor snags in the second fix joinery or plasterwork, especially if he is going to be the one to decorate the rooms.

The main alternative to a dormer type of window is the Velux type of openable roof light. This is looked at in detail in Project 10.

Loft conversion - 3

In this Project we complete the conversion by dealing with the dormer window and all the applied finishings etc.

		£
a)	Provide, erect, maintain, dismantle and clear independent scaffold for roof works,	330.00

b)	Carefully remove roof tiles over area of dormer and set aside for re-use,	
	Two men: .33 day	33.15
c)	Provide temporary weather covering to exposed portion of roof area,	220.00
d)	Remove rafters in area of dormer and set aside for re-use including providing any necessary temporary supports,	
	Two men: .25 day	25.11
e)	Supply and fix two new trimming rafters,	
	Two men: .125 day	12.56
	Material:	30.00
f)	Construct dormer window approximate size 2.50 x 1.00 metres high including new head and cill trimmers, windowhead, cheek studding and rafters,	
	Two men:two days	200.92
	Material: (assumed use of removed rafters plus new where necessary,)	44.00
g)	Supply and fix suitable softwood window complete with Insulight or similar double glazing,	
	Two men: .33 day	33.15
	Material:	215.00
h)	Supply and fix external quality ply to dormer cheeks,	
	Two men: .25 day	25.11
	Material:	13.00
i)	Provide and fix new felt and battens and re-fix previously removed tiles and supply any replacement or additional tiles required,	
	Two men: half day	50.23
	Materials:	33.00
j)	Provide and fix lead sheet to dormer cheeks including underfelt, flashings, and all labour required (approx. 3,00 SM)	150.00
k)	Supply and fix UPVC gutter and drainpipe including all fittings on and including softwood fascia and soffit board,	
	One man: .75 day	37.67
	Material:	40.00
l)	Line soffit and internal cheeks of dormer with insulation quilt and plasterboard for direct decoration,	
	Two men: half day	50.23
	Material:	66.00
m)	Allow for making good any damage to existing roof, mastic pointing and general sundries externally,	
	One man: half day	25.11
	Material:	22.00
n)	Prepare and paint all new exposed external woodwork etc,	
	One man: .33 day	16.58
	Material:	6.00

o)	Supply and fix window board,	
	One man: .25 day	12.56
	Material:	8.00
p)	Supply and fix skirting and architraves and mouldings generally,	
	One man: one day	50.23
	Material:	43.00
q)	Supply and fix one room door including lining, stops and all ironmongery and four cupboard doors (1500 x 838mm)	
	Two men: one day	100.46
	Materials:	250.00
r)	Allow provisional sum for all internal decorations consisting of lining walls and ceiling, emulsion to ceiling, vinyl paper to walls and prime and three coats to woodwork including new staircase and balustrade,	650.00
s)	Allow to clean down and leave ready for occupation,	
	One man: one day	50.23
		2703.30
t)	Allow for prelims 12½%	337.91
		3041.21

It should be noted that Projects 7, 8 and 9 dealing with Loft Conversion have dealt with a very basic and straightforward specification. Due allowance must be made for any of the other matters mentioned in the text.
It should also be noted that on work of this nature supervision is essential and time will be taken to produce drawings, planning applications etc. due allowance for which should be included in the estimate.

Project 10
Installation of roof windows

Most of the advantages, disadvantages and problems associated with loft conversions have been dealt with in the last three projects. The openable roof window is a very practical and cost-effective way of introducing light and ventilation into a converted roof space but there are quite a number of problems that will be encountered even with this simple job.

As the majority of the work concerning a roof window can be carried out from inside the building there is not normally a need to provide a scaffold. Scaffolding can be very expensive, so if the roof in question is reasonably accessible from ladders it is suggested that a scaffold is not used.

The roof tiles or slates in the vicinity of the proposed window should be carefully removed and laid aside for partial refixing on completion of the job. The rafters can then be trimmed to suit the size of the new roof window. This is one part of the job that will probably require building regulation permission or approval under the London Building Acts as the actual structure of the building is being altered. It is usually sufficient to trim the opening with timbers of a similar size section in order to replace the strength and stability that has been lost with the removal of part of the rafters.

The installation of the roof window will almost certainly be carried out in conjuction with a loft conversion of some sort and in this respect the trimming of the rafters to insert the roof window will only be part of the building regulations application.

Unlike a dormer window, a roof window will not increase the

headroom to any part of the attic. In a roof with a pitch of 45 deg. or more this will probably make no difference to the overall headroom but on shallower pitches headroom plays a far more important part. 2.3 metres is the usual minimum headroom but as long as there is the same floor area, with a headroom of 1.5 m or more, the Building Regulations requirements will be met.

Sometimes the only reason for installing a roof window is to allow extra light into a room that will be used as a store. Headroom will not then be a problem as this will not be considered a habitable room.

Planning permission is not normally required when a roof window is being installed, although this will depend very much on the

The Velux type GML double function roof window in the escape opening position.

Roof window shown in Myresjö timber frame home.

building involved. Listed buildings always require planning permission for a change in appearance to the external elevations. A planning application may also be required if the room will increase the volume of the property by 70 m^3 or more. This Project is only a general guide and it is suggested that the client contacts the local authority to determine whether or not an application is required.

When estimating for the installation of a roof window it is likely that the job will be carried out as part of an overall conversion and as such the other provisions, e.g. plaster boarding out the roof space, insulation, are likely to also provide part of the overall estimate. In this respect only the roof light itself is dealt with.

The pitch of the roof will decide the angle of the roof window but

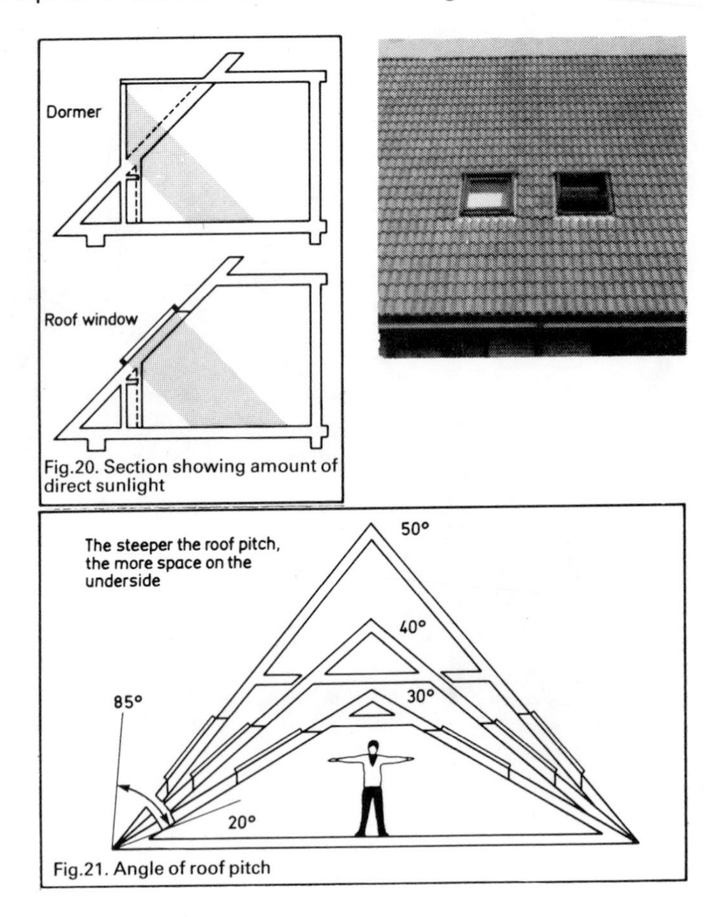

Fig.20. Section showing amount of direct sunlight

Fig.21. Angle of roof pitch

nevertheless in most cases a roof window will provide more sunlight than a dormer window would. This can be seen in Fig.20. The only problem with direct sunlight is that it increases the thermal gain in the room and in the summer months, combined with the insulation between the rafters, can make the room uncomfortably hot. Many manufacturers of roof windows provide blinds either as extras or as integral parts of the window. Another way of controlling the thermal gain is by applying a solar reflective film to the underside of the window. Only about 2 per cent or 3 per cent of the daylight is lost, but almost 95 per cent of the heat-gain can be eliminated. The quality of these solar reflective films has

Above: Roto 320 laminated pine frame roof window from Magnet and Southerns. Top right: Orion roof window from Ubbink blends with most types and shades of roof tile. Its integral flashing simplifies installation.

improved over the past few years and whereas they used to bubble and peel away from the windows, they now appear to be almost part of the window itself.

Another recent innovation is the Pilkington Insulight glazing unit. This is basically an insulated form of glazing which will help reduce heat loss in the attic room.

The height of the roof window in the room should also be considered. A clear view is normally required whether the householder is standing or sitting in the room, but if this is not possible, and the roof light has to be positioned at a height above normal reach then a remote control opening and shutting operation may be necessary. Velux manufactures a system of cords and pulleys to open its windows as well as an electrically-controlled opening mechanism.

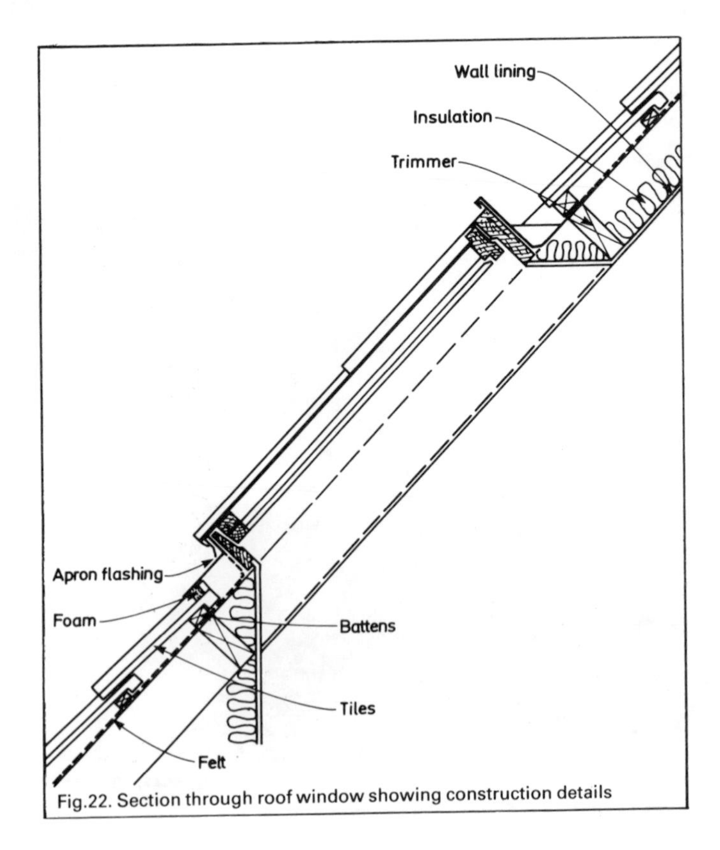

Fig.22. Section through roof window showing construction details

As the whole of the glazed area is normally openable ventilation can be greatly improved.

The window should be installed in accordance with the manufacturer's instructions which should prove to be relatively straightforward. The roof window should be properly weathered in its new opening. Velux manufactures preformed flashings to make this easier.

Velux produces a good booklet entitled *Attack The Attic* and it is recommended that the client obtain one of these booklets to enable him to choose the type and style of roof window and to consider all the different possibilities.

Roof windows are a straightforward, cost-effective way of improving light and ventilation to a roof space and as long as all the local authority requirements and manufacturer's instructions are met, then there is no reason why this job cannot be carried out both swiftly and effectively.

Installation of roof windows (size 940 x 1600mm)

It has been assumed that all internal finishings have been dealt with elsewhere. Should this not be the case allowance should be made for making good any internal finishings disturbed or damaged as a consequence of the installation.

		£
a)	Provide tower scaffold to gain access to roof for renewal of tiles and re-fixing etc.,	
	Labour: two men .25 day	25.11
	Plant:	52.00
b)	Carefully remove tiles from roof and set aside for re-use to an area allowing adequate space to form the required opening	
	Labour: two men .33 hour	4.25
c)	Support rafters as necessary, form opening providing and fixing trimmers at head and cill of opening,	
	Labour: one man .50 day	25.11
	Material: say	5.00
d)	Supply and fix Velux roof light type GGL3 complete with flashings,	
	Labour: two men .33 day	33.15
	Material: say	190.00
e)	Make good roof felt, re-fix battens and tiles and leave watertight,	
	Labour: two men .25 day	25.11
	Material: say	3.00

f)	Remove scaffold, clear away surplus tiles etc.,	
	Labour: one man one hour	6.44
	Transport: say	11.00
		380.17
g)	Allow for prelims 12½%	47.52
		427.69

The foregoing labour rates are based on £4.60 per hour plus 40% for profit and overheads.
Both the hourly rate and percentage should be adjusted as necessary.

Project 11
Installing secondary glazing

Many householders do not realise that the distance between the original glazing and the secondary glazing determines whether the room is going to be insulated for acoustic or for thermal reasons.

There is also a difference between double and secondary glazing and this should be explained to the householder.

Double glazing is normally manufactured from a hermetically sealed unit, glazed in a purpose-made frame with an extra large rebate in the usual manner with sprigs and putties. Normally when a house is double glazed it is done either at the time of construction or when the windows and frames are completely replaced. The distance between the two pieces of glass is usually very small, i.e. no more than about 10 mm, and therefore helps to increase the thermal insulation.

Secondary glazing however, would be carried out for completely different reasons and in a completely different situation. With the original window in place the new glazing is installed usually some distance away from the original. If it is carried out for sound insulation the squares of glass should be approximately 100 mm away from each other.

If the glazing is carried out for thermal requirements it is pointless just re-glazing the windows with double glazed units, even if the rebates are large enough, because the opening casements of the window may be draughty.

Sometimes both types of glazing are carried out to provide both thermal and sound insulation.

As this project deals with secondary glazing the suitability and condition of the original glazing will not be discussed. It should be taken into consideration, however, when an estimate is being prepared and the client advised on the various solutions to his problem.

Secondary glazing can be carried out completely from the inside of the property and the householder should be advised that almost every room in his property is likely to be in a state of disarray during some stage of the job.

Secondary glazing can be either sliding or side hung. Both have their various advantages and disadvantages although with sliding secondary glazing the amount of natural ventilation allowed into the room is more controllable than with the side hung type.

The frames to support the glass can be made of aluminium, timber or plastic. This depends largely on the property and on the requirements of the owner. If the existing windows are made of painted softwood, satin anodised aluminium may look rather out of place. If the areas of glass to be installed are large they will have to be supported in something more substantial than just a slim section of plastic.

Polycell manufactures a do-it-yourself secondary glazing kit which consists basically of a plastic edging strip, to hold the glass in place and protect the edges, and various clips and brackets to fix the glass in position. This is fine for small windows and a do-it-yourself man but is not always the most suitable method of providing secondary glazing . One of the main problems with the Polycell system is that if the job is being carried out on a tight budget it is likely that the squares of glass will be permanently fixed and ventilation will not be available once the glass is in place.

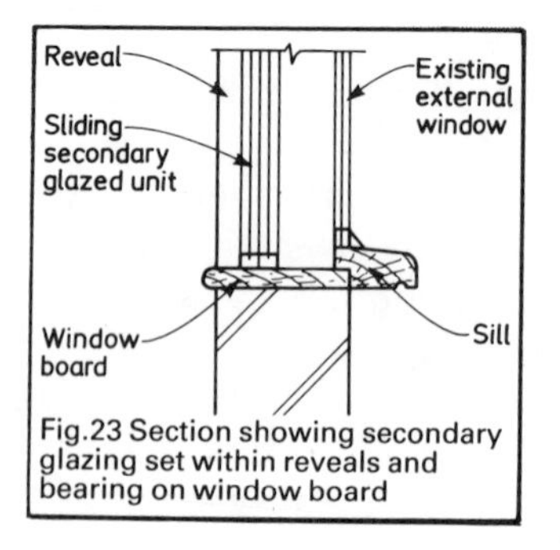

Fig.23 Section showing secondary glazing set within reveals and bearing on window board

Polycell does manufacture openable squares but once again the weight of the glass and the cost effectiveness should be taken into consideration.

This system can be totally removed in the summer months and stored in the garage or attic.

Many secondary glazing contracts are carried out under government funded road schemes. This means that the grant is made available to householders who live near to motorways, etc., and as such the secondary glazing has to conform to certain government requirements. The reveals, head and sill between the old and new glazing should be insulated with a proprietary material to deaden the sound between the two squares of glass. If the job is being carried out under such a scheme all requirements will need to be fulfilled.

The secondary glazing should be properly sealed at all edges to ensure that there are no air gaps. Sound will penetrate through even the smallest gap, reducing the effectiveness of the secondary glazing.

If the reveals and the existing are not square the new glazing will either have to take this into account or will need to be properly packed out at the edges to compensate.

If secondary glazing is installed to prevent the passage of sound it is not normally opened. A mechanical extract may then need to be installed in the room to compensate for the lack of ventilation. The only problem with this, however, is that the mechanical vent may prove to be just as noisy as the road or air traffic noise that the secondary glazing is installed to combat!

Curtains and curtain rails may need to be repositioned and this should be allowed for in the estimate.

Secondary double glazing is a good way of providing additional security to a property and while the windows are being installed the type of fastening should be determined with the client.

The sight of secondary glazing in a property can be enough to deter the majority of intruders and burglars.

Blinds may be required to combat the heat gain in the room and the problems associated with the fixing of these should be thought about prior to the estimate being prepared.

Secondary double glazing can also be fitted to the outside of the window reveals in certain circumstances, but this is not advisable as the weight of the glass is not properly supported.

Installation of secondary glazing

As described above, it is normal practice these days to employ a specialist firm to supply and fix complete, however there are occasions when during the course of other work being carried out to a property one or more windows are required to have secondary glazing and in these circumstances the builder may well carry out the work himself. As a guide we have assumed a window 2.00 metres wide and 1.50 metres high, hardwood surround and glazing sliding in aluminium channels.

		£
a)	Check opening for actual size including diagonals for squareness; remove any obstructing mouldings and increase depth of window board as required,	
	Labour: one man .25 day	12.56
	Materials: say	7.00
b)	Provide and fix hardwood sub-frame including plugging and screwing to head and reveals and bedding in mastic to seal all gaps where surfaces are not true or flat,	
	Labour: one man .50 day	25.11
	Material: say	42.00
c)	Provide twin groove aluminium channel and fix to hardwood sub-frame,	
	Labour: one man .125 day	6.28
	Material: say	22.00
d)	Supply felt draft strips and fix to glass or aluminium,	
	Labour: one man one hour	6.44
	Material: say	7.00
e)	Supply aluminium window pulls and fix to glass with adhesive,	
	Labour: one man .33 hour	2.15
	Material: say	5.00
f)	Supply polished plate glass sliding panels and set in position in aluminium channel (say 3 No)	
	Labour: one man .25 hour	1.61
	Material: say	95.00
		232.15
g)	Allow for prelims 12½%	29.02
		261.17

The foregoing labour rates have been based on £4.60 per hour plus 40% for profit and overheads.
Both the hourly rate and percentage addition should be varied as necessary.

Project 12
Building a small extension

The reasons for building an extension are numerous. The family may be increasing and an additional bedroom may be required, an extra reception room, larger kitchen or just more living space. Whatever the reasons for building an extension it is important that it is well planned, well built and from an appearance point of view matches the existing property.

At today's prices a rough estimate of a traditionally built extension would be approximately £60.00 per sq. ft. (in the South of England) although this amount would alter, depending upon the size and location of the extension. This is a very rough guide and the client should be made aware that building costs, especially for work such as this, are certainly not cheap. It is therefore extremely important to make the most of the extension and the client's reason for wanting to extend his house should also be determined. If his money is rather tight there may be a more cost effective way of achieving what he requires without having to extend at all.

Usually, for this type of work, an architect or surveyor would be involved to produce various plans and submit applications to the local authorities. Planning permission would probably be required but if the proposed extension is less than70 m^3 in volume (or 15 per cent of the existing house whichever is the smaller) then this would be known as a permitted development.

This does not necessarily mean however that planning permission would not be required. A building may protrude beyond the building line or be too close to a boundary. It may block out the adjoining owners daylight or simply be aesthetically unbalanced.

In any case the plans sould be submitted to the local authority and if you are not sure whether or not planning permission is required, ask the planning office.

It is good practice to ensure that the client informs his neighbours of his proposals and asks them if they have any objections at all. Far better to iron out any problems in the very early stages of the project than upset the neighbours and waste both time and money. A large extension to a terraced house, even if it is only a single storey one, is likely to produce various reactions from the adjoining owners and their objections may be well founded.

The building may be a listed one and therefore discussions with the local authority's Listed Building Architect are important, as there will almost certainly be restrictions as to the type of extension that would be allowable and the type of materials that should be used.

The subsoil should be investigated, the depth of the existing foundations ascertained and the size, depth and type of the new foundations should be calculated to ensure that the possibility of differential settlement is minimised. Foundations should of course be designed to suit the type of subsoil and in extreme cases specialist tests may need to be carried out.

The site may well be constrained by the positioning of trees and drainage and these too should be investigated. If any trees are to be felled the local authority may have a say in this. The felling of trees needs to be mentioned in the planning application and it may be that the local authority will require new trees to be planted elsewhere in their place.

It is advisable that a thorough drainage test and report is carried out by a specialist prior to any works beginning, as there is little or no point in connecting new drains to existing defective ones. Far better to renew, replace or even rerun any defective drainwork during the course of the other groundworks.

One point that will almost certainly affect the estimate is that of access to the site. If the new extension is to be at the back and the necessary machinery cannot be made use of then the work will have to be done by hand. This will certainly increase the cost of the job and may be further hampered by the advent of poor weather conditions. Hand digging of deep foundations is a long and laborious task and any deep foundation trenches may need to be planked and strutted in order to comply with the Construction

Regulations. Ground works can eat into quite a large portion of the budget and a lot of money can be spent before the extension has even risen from the ground.

Another problem that would be worsened by poor access is the moving of materials and debris. If sand, cement and bricks, etc., all have to be wheelbarrowed some 50 yards or so before they are in a position to be used, the builder will have a labourer doing almost nothing else for a couple of weeks.

There are obviously many sections of the Building Regulations (or London Building Acts) that will have to be adhered to or complied with in regard to the groundworks. As mentioned above, specialist calculations and reports may be required for the foundations and it should be made clear that unforeseen works and additional depths of foundations, etc., will have to be charged as an extra, usually on a day work basis. If you are working to a contract this would normally be the JCT Minor Works Contract and the builder should satisfy himself that he knows exactly what he is letting himself in for when he signs this contract.

Many extensions have to built over existing drain runs, and while it is not always necessary to renew the drainage pipes in cast iron where they pass under the external walls of the building, it is necessary to bridge the drain runs in some way. This is usually achieved with a couple of precast concrete lintels or whatever is necessary to support that section of the wall. It is unnecessary to go into too much detail about the various Building Regulations, but these should be incorporated into the drawings and specification by the architect or surveyor. Should the client wish to proceed without the professional assistance of an architect or surveyor the builder should let him know that there are various planning and construction laws that he will need to comply with, and that by far the simplest way of doing this is to employ a professional. It would also be nigh on impossible for a builder to accurately price for the building of an extension and to carry out the works without some form of specification and drawings to guide him.

Another point that should be made clear to the client, is that the foundations for a single storey extension will not necessarily be adequate for a two-storey extension should the owner ever wish to carry out such a project.

With regard to the drainage, it may be necessary for an existing external manhole to become an internal one. An internal manhole would need to be fitted with a double seal cover and frame which can be screwed down to reduce the possibility of the drainage backing up and sewage entering the house. If there is an external soil pipe running down the wall of the building this may too need to become an internal pipe.

There is a high possibility that the existing pipe will be of the old cast iron type and in this case it is advised that it be renewed in pvc to ensure that there are adequate removable rodding eyes for access to the pipe to clear any blockages that ever occur. The pipe would then need to be boxed in and it is advisable that if this is the case the new internal pipe be positioned in a kitchen rather than a living room. Ensure that the boxing itself can be removed, otherwise there would be no point in ensuring that the pipe can be rodded.

A new manhole may need to be built if the drainage is to alter to any great degree and this may well be necessary if the new extension is to be used for a kitchen. Figs. 24 and 25 show a cut away section through a new manhole and a plan.

In order to construct a new extension it may be necessary to

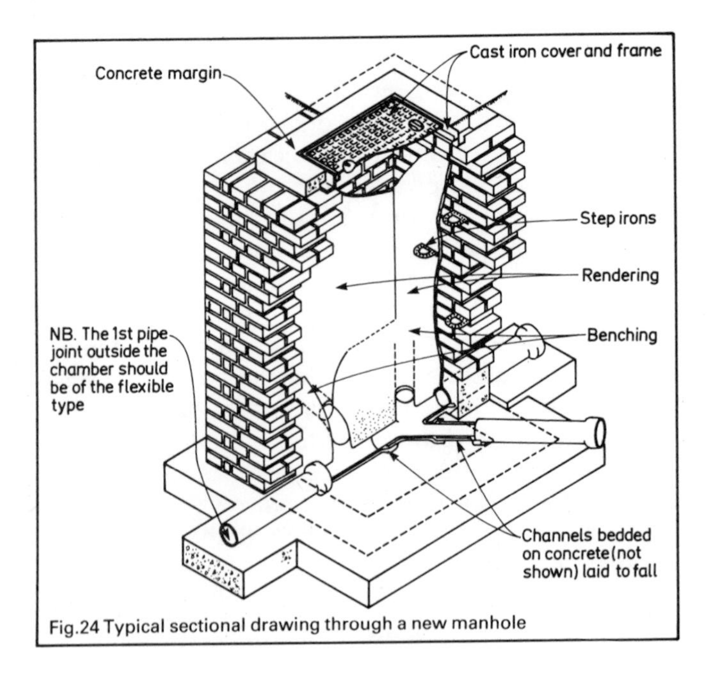

Fig.24 Typical sectional drawing through a new manhole

excavate to a considerable depth very close to the existing building. This may undermine the existing foundations and temporary support may be required. It is important to try to determine the type of existing foundations and the subsoil.

Safety should be of prime concern, not only of the operatives on site but also the safety of the building. No contractor, especially a small one, wishes to have a large insurance claim made against him.

Finally it may be the case that the existing ground floor is of suspended timber construction and that the new extension is to be built with a solid ground floor. It is important therefore that the ventilation to the suspended timber floor is not impaired in any way. It may be that some additional ventilation grilles can be installed around the property or that a plastic waste pipe will need to be bedded into the concrete to allow the timber floor to be ventilated through the solid floor to the outside air.

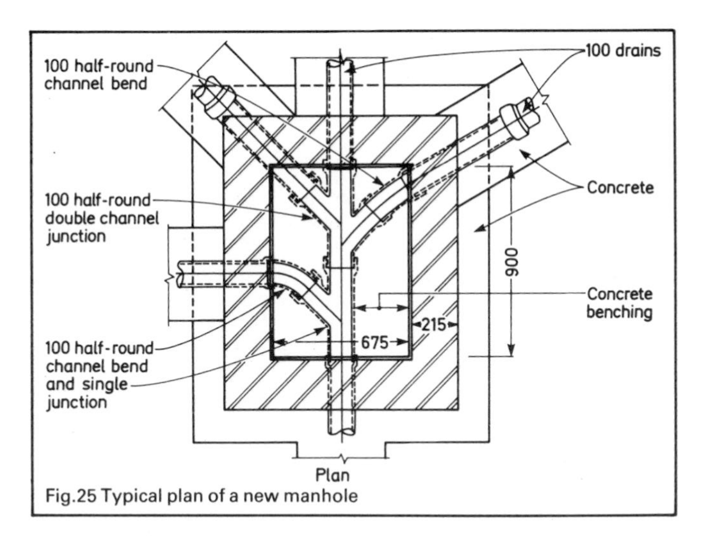

Fig.25 Typical plan of a new manhole

Domestic Extention Part 1

Note: The following estimated costs refer to a single storey extension to an existing brick built dwelling on a level site with side access for removal of waste and entry of new materials.
An internal size of 5.50 x 3.00 x 2.45 metres has been assumed.
Suitable adjustments to time and cost will be required to allow for sloping sites, hard digging, bad access, inaccessible drains etc. A careful site visit is essential to

ascertain any problems that may arise. It is also necessary to establish the order of construction and allowable working hours as these may be determined by the client, particularly where old or infirm people are involved. Cost adjustments will be required to cover any or all of these points.

Works to dpc including drainage:

		£
a)	Allow to clear area of extension of all paving and general obstructions and set out (assume precast paving on lime mortar bed) and excavate oversite;	
	Labour: 2 men .75 day	75.35
b)	Excavate foundation trenches, bottom up, remove surplus soil. (It may be possible to spread and level surplus soil in the garden area - we have allowed for loading into skip and carting away). (Exc. 7 cm cart away 4 cm);	
	Labour: 2 men 2.50 days	251.15
	Plant: skip	40.00
c)	Concrete in foundations (2 cm);	
	Labour (including wheeling in): 2 men .5 day	50.23
	Material: ready mix concrete	100.00
d)	100mm bed hardcore and blind with sand (allow for compaction)	
	Labour (including wheeling in): 2men .25 day	25.11
	Material: (2 cm) hardcore	25.00
	Material: (1 cm) sand	18.00

Before either (c) or (d) can be carried out it will probably be necessary for the Building Inspector to examine and pass the excavation; due allowance should be made for any standing time involved.

		£
e)	Damp proof membrane laid oversite carried up face of existing wall, tuck into joint including raking out and point in cement and sand; leave surplus at all edges to build in to new walls;	
	Labour: 1 man .33 day	16.58
	Material:	7.50
f)	Concrete bed oversite 100mm thick;	
	Labour (including wheeling in): 2 men .5 day	50.23
	Material: (2 cm)	100.00
g)	Brick cavity wall to dpc level including cut tooth and bond to existing walls;	
	Labour: 1 bricklayer 1 labourer 2.50 days	251.15
	Material: Bricks (PC £110 per 1000) 1200	132.00
	Material: Mortar .40 cm	19.20
h)	Damp proof course;	
	Labour: 1 bricklayer 1 labourer .125 day	12.56
	Material:	15.00
i)	Backfill around foundations (3 cm) including excavate from spoil heap and wheel;	
	Labour: 2 men .25 day	25.11

j)	Depending on the nature of the ground it may be necessary to timber the trenches. Allow a provisional sum for this work including any necessary extra excavation and backfill consequently required,	150.00
		1363.96
k)	Allow for preliminary abd 12½%	170.50
		1534.46

Drainage

As no internal plumbing requirements are involved we have only dealt with rainwater disposal and have assumed an existing manhole is available to connect into

a)	Excavate trench for drain including subsequent backfill and cart away of surplus;	
	Labour: 2 men .50 day	50.23
	Plant: skip	7.00
b)	Break into existing manhole for and make new connection including making good brickwork and benching;	
	Labour: 2 men 1.75 day	175.80
	Material:	15.00
C)	Lay and joint 100mm drain (3 LM);	
	Labour: 1 man .33 day	16.58
	Material:	11.00
d)	Trapped gulley surrounded with concrete;	
	Labour: 1 man .33 day	16.58
	Material:	16.00
e)	Bed and surround pipe in shingle;	
	Labour: 1 man .33 day	16.58
	Material: (.75 cm)	12.00
		336.77
f)	Allow for prelims 12½%	42.10
		378.87

The foregoing labour rates are based on £4.60 per hour for a 39 hour week plus 40% for profit and overheads; adjustments should be made accordingly for any variation required to these.

Collection

Substructure	1534.46
Drains	378.87
	1913.33

Project 13
Roofing and extension

The superstructure of the new extension will depend largely on what the client wishes to use the new building for and what the appearance of the existing building is. It may be that he only wishes to build a fully glazed conservatory to use as a sun room and to help grow his tomato plants! A completely different form of construction above ground would then be necessary, although the ground works as mentioned in the previous project would still remain more or less the same.

If a conservatory was required, there are many firms around at the moment who supply and/or erect factory-built conservatories, so the builder would only need to construct a slab and mainly carry out some internal works like forming a new opening through the existing wall.

Nothing was mentioned in the last project about damp proof courses and damp proof membranes, although it is important to include a dpm in the ground floor slab and this should of course tie into the dpc of the existing property. The dpc should also be installed into the new external walls and this too should tie in with the existing, so that there is no possibility of dampness penetrating into the inside of the house or rising in any way.

If the extension is to be constructed from brick then the new brick should match the existing as near as possible. If it is not possible to obtain bricks with a near likeness it is suggested that the new construction should be either rendered or a completely contrasting brick obtained. Nothing looks worse than an extension that is obviously an extension. Pointing to the brickwork should also match the existing not only in the colour and type but also in the way it is finished, i.e. flushed joints, raked joints, weather struck joints, etc.

Many buildings that are extended are of the 1930s type, with brickwork up to the sill of the ground floor windows and rendering above. This too should be matched if at all possible, although the client should be made aware that rendering, brickwork, pebble dashing, etc., built to exactly match the existing house will increase the cost of the building work. He should also be made aware that nicely built and properly blended-in extensions will increase the value of a house far more than shoddily built and aesthetically incorrect extensions will.

External walls will have to meet far more stringent thermal insulation values than their counterparts had to in previous years.

"Timber frame is now an acceptable form of construction."

Most blockwork manufacturers now produce comprehensive technical literature and there is no excuse for not knowing what the insulation value of any particular type of construction will be.

It may be that for the sake of his finances the client wishes to use timber frame construction. This is an acceptable form of construction nowadays especially to the local authorities. There is still quite a lot of market resistance to timber frame construction as many people believe that it is inferior to blockwork. Apart from poor site practice there is no reason to believe that timber frame construction should last for any shorter length of time than a traditionally built extension. It is more common to use this form of construction for the building of a whole new house rather than an extension, although no doubt the majority of timber frame manufacturers would be pleased to estimate for the cost of the timber frame kit for an extension.

The superstructure should be properly bonded into the existing and although 'blockbonding' may be acceptable for the internal skin of blockwork, the external skin of brickwork should be properly toothed and bonded in order to adequately tie in the new to the old.

Windows too should match existing not only in their construction but also in the materials and the height to which they are installed. The majority of people wish to completely redecorate the exterior of their house when an extension has been built and this is highly recommended if there is any rendering involved.

Brick arches are seldom used nowadays in the construction of external window and door openings and in almost every instance where an opening is required in new work it is recommended that a Catnic type of steel lintel be used. This is now acceptable to all local authorities in the outer metropolitan areas and the majority of district surveyors in the London area are also accepting these lintels in external openings.

The type of roof construction that will be used over the new extension will depend on a number of things. If the roofs to the existing property are of pitched construction it is highly recommended that a pitched roof be installed over the new extension. This is not always entirely possible though, as on a single storey extension a pitched roof may rise above the sill height of any first floor windows. It is not usually an insurmountable problem however and for the sake of the look of the building and its resale value it is certainly best to install a pitched roof. If the extension is two-storey there is even more reason to install a pitched roof, properly built into the existing roof, even though this may increase the cost of the job considerably.

It may be the case that the client wishes to use the flat roof of a single storey extension as a patio accessible from first floor level. If the roof is going to be subject to such traffic the ceiling joists will

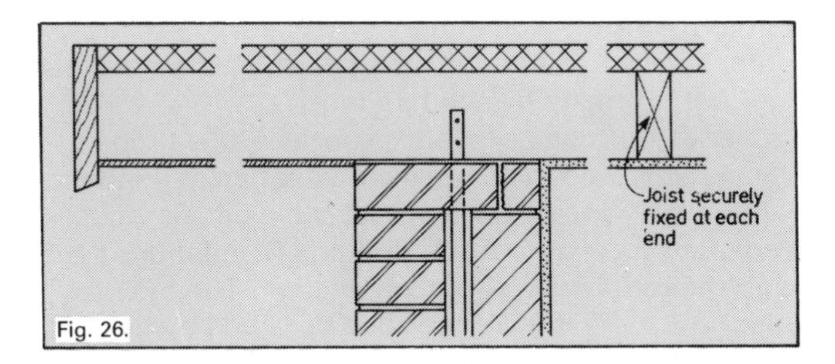

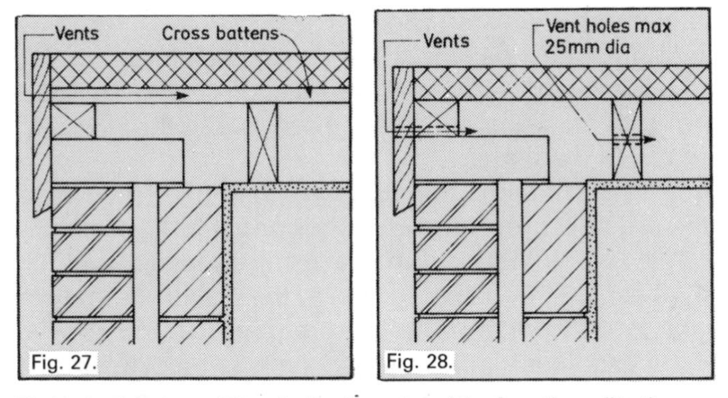

Fig.26. Anchorage could be obtained by external ties from the wall to the underside of joists or some form of strap fixing along the joists and down the cavity might be devised. For a substantial overhang, the strap fixing should be anchored well down the walling.
Fig.27. Whenever possible it is better to provide any required roof ventilation in the direction of the roof joists. If this is not possible, cross battening is necessary for good ventilation; this also has the advantage of providing for falls at right angles to joists when required.
Fig.28. Small holes bored at mid-depth of roof joists are sometimes used for ventilation. Where roof materials are not vulnerable to damage by dampness and conditions require only minor ventilation this method may be adequate. It should not be regarded as suitable where the situation requires a fully ventilated roof.

have to be designed to cope with this load and the roof covering will need to be of a sufficiently high quality (e.g. asphalt).

A cheaper form of pitched construction which would look all right without necessarily costing the earth would be a lean-to roof. This is a reasonably good compromise that can, if designed correctly, look every bit as good as a pitched roof.

It must be stressed that it is very important to match the new roof tiles or slates with the existing, especially if the roof is to tie into the existing roof. One way of achieving this is to strip some of the tiles from the rear of the property and use them on the new extension at the front, so that they will blend in exactly. Any new tiles that then have to be bought can be used on the rear elevation, which will not normally be seen so much.

The general roof construction, especially the timbers and the eaves details, will need to be calculated bearing in mind the type of roof covering that is to be used and all such details will have to be submitted to the local authority's Building Regulations department along with the remainder of the application. The majority of local authorities now insist that Code 5 lead be used as a flashing where the new building abuts the old and the stepped flashings will have to be chased into the existing brick courses to ensure that there is no possibility of rainwater penetration.

If conditions and finances make it impossible to use a pitched roof and a flat roof does have to be used, the Building Regulations are quite strict on the various construction details. Figs. 26, 27 and 28 show eaves details.

It is important to allow adequate ventilation to the roof timbers and there are now proprietary meshes and ventilators, etc., on the market that can be installed in the soffit at eaves level. It will be normally best in this case to close the cavity. It may also be necessary to drill small holes at mid-depth in the roof joists to achieve the ventilation in a situation such as that shown in Fig. 28, where the joists are running parallel to the external wall and it is not possible to achieve adequate ventilation in any other way.

It is sometimes necessary to install roof lights in a single storey flat roofed extension to improve the natural daylighting in a room that may now have become internal because of the building of the extension. There are many proprietary brands of roof light on the market and these can be made quite an attractive feature from the inside.

Domestic Extension Part 2

This project deals with the superstructure and follows on from project 12. The construction has been assumed as cavity wall of facing bricks and blocks with cavity insulation and plastered internally; timber roof with insulated deck and three-layer built up felt roof. Floor screeded with vinyl tiles. One softwood window to flank and patio doors to garden.

		£
a)	External wall in facing bricks (PC £200 per 1000), inner skin 100mm Thermalite or similar blocks with Vencil Resil or similar cavity insulation;	
	Labour: One bricklayer and one labourer (23 SM) 7 days	703.22
	Wheel in; 1 man 1 day,	50.23
	Material: Bricks 1500 @ £200.00	300.00
	Blocks 23 SM @ £5.00	115.00
	Mortar 1 CM @ £40.00	40.00
	Insulation 23 SM @ £2.60	59.80
	Steel lintels	66.00
	Damp course	8.00
	Wall tiles & sundries, say	20.00
b)	Cut tooth and bond and make good to existing wall;	
	Labour: 1 bricklayer & 1 labourer 1 day	100.46
	Materials: say,	16.00
c)	Timber wall plate and roof joists including hangers,	
	Labour: 2 men x 1 day	100.46
	Material: Wall plate	19.50
	Joists	80.00
	Hangers	15.00
	Sundries	15.00
	Strutting	20.00
d)	Roof decking and felt roofing	
	Labour: 2 men 1.50 days	150.69
	Material: Pre-screeded wood cemair decking,	97.50
	Nails and sundries	10.00
	Insulation	230.00
	Felt roofing including all flashings etc. 20 SM @ £32 by sub-contractor	640.00
e)	Fascia and Soffit boards;	
	Labour: 2 men .75 day	75.35
	Material:	52.00
f)	Rainwater gutter and downpipe in uPVC;	
	Labour: 2 men .50 day	50.23
	Material:	20.00
g)	Windows and doors; 340WT window and MPD8 patio door in softwood frame including double glazing and closing cavity at reveals;	
	Labour: 1 bricklayer and 1 labourer 1 day	100.46

	Material: Window including glass	114.00
	Patio doors including glass	385.00
	Sundries including mastic	20.00
h)	Plasterboard and set to soffit and render and set to walls including rake out joints of existing wall;	
	Sub-contr.: Ceiling 17 SM £7.	119.00
	Walls 38 SM £8.	304.oo
	Labour: Rake out; 1 man say 1 day	50.23
i)	Screed and vinyl tiling to floor;	
	Sub-contr.: Screed 17 SM @ £8.50	144.50
	Vinyl 17 SM @ £6	102.00
j)	Allow for window boards and skirtings;	
	Labour: 1 man .50 day	25.11
	Material:	30.00
k)	External decorations to window, patio door frame, fascia and soffit;	
	Labour: 1 man 1 day	50.23
	Material: say	6.00
l)	Internal decorations; emulsion to walls and soffit, paint to window and skirtings etc;	
	Labour: 1 man 3 days	150.69
	Material: say	20.00
		4695.66
m)	Allow for prelims 12½%	586.96
		5282.62

The foregoing labour rates are based on £4.60 per hour for a 39 hour week plus 40% for profit and overheads; adjustments should be made accordingly for any variation required to these.

Project 14
Completing an extension

Apart from the basic construction of the new extension there will be other matters to consider, especially regarding the services. Electricity will need to be taken into the new extension and various socket outlets and lighting, etc., will need to be provided. The main fuse board to the property will hopefully have a spare fuse that can be used to run a new ring for the power. In most circumstances it is not recommended that the existing ground floor ring be spurred off unless there are only a couple of socket outlets to be provided. If the extension is only going to be used as a living room this may be the case but if it is going to be a kitchen there are likely to be dishwashers, washing machines, and maybe even a cooker that will need its own 35 amp fuse.

The heating to the new extension could cause more of a problem. If the existing house is centrally heated the boiler is likely to be adequate for one or maybe two additional radiators but this should be checked. Don't forget that the extension is likely to have three external walls and so the radiators should be properly sized to compensate for this. They should also match the existing and if the pipework to the radiators is to be installed in the screed of a solid floor it should first be lagged. The lagging is used not so much to provide thermal insulation as to provide a barrier between the copper piping and the corrosive salts in the screed.

If wastes are to be fitted into an existing soil stack and this stack is constructed on the single stack system it is important that the wastes are fitted in certain positions and heights in the stack to comply with the Building Regulations. Problems with the plumbing can occur if these rules are not adhered to and it is advisable to employ a competent plumber to carry out such work.

As mentioned in Project 13, the external wall will have to reach the present day thermal insulation requirements and this also applies to the roof space. One hundred millimetres of a glass fibre insulation quilt laid between the ceiling joists will be adequate in pitched roof construction although it may be more feasible to provide a different form of insulation such as Jablite to a flat roof.

Another item of construction that will need to comply with the Building Regulations is the style of window used in the external walls. There are maximum and minimum window areas laid down in the regulations in order to control the amount of natural ventilation, daylight and thermal insulation for a room. Normally common sense will tell you when a window is right or wrong although window area calculations may be required by the local authority as part of the Building Regulations Application. Any internal rooms will require some form of mechanical ventilation, although this will normally only apply to bathrooms and toilets.

Naturally the floor levels in the new extension should match those in the existing property at ground level, although it may not always be possible to match the floor levels if the extension is to be taken to two storeys in height. The minimum ceiling height nowadays is 2300 mm. If the property is a sixteenth century farm cottage, it is likely that the ground floor ceiling height will be lower than the required 2300 mm and if so the first floor level in the extension will have to be higher than the first floor level in the existing house.

Some houses are extended because they are the traditional

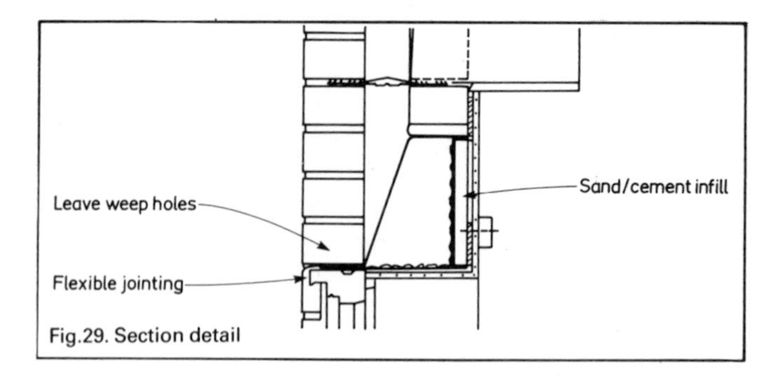

Fig.29. Section detail

workman's cottages without a bathroom. In this instance the client may wish to provide a bathroom at ground floor level adjacent to the existing kitchen. Again there are certain regulations that must be adhered to if this is to be carried out, the most important of which is that a ventilated lobby will need to be provided between the bathroom and the kitchen.

Rainwater on the new extension will have to be disposed of. If the existing property has a soakaway it may be that this will be too near to the new extension and a new one will have to be dug. This would be an expensive item and one which should be taken into consideration when the estimate is being prepared.

Some people may just wish to convert the old external door from the existing building into an internal door and gain access to the new extension this way. The majority of people, however, would probably knock an enlarged opening into the old external wall. Structural calculations will need to be provided and submitted to the council to show the suitability of any lintel or universal beam and the beam would need to bear onto suitable piers and be encased to a half hour fire resisting standard. Temporary support will be required when the opening is being formed (refer to Project 3 about forming an opening between rooms).

This part of the work is usually best left until near the end of the job so that the dust and dirt and the weather will not enter the house.

During the job there are likely to be certain areas of unevenness, etc., to the walls and these can normally be disguised by a competent plasterer.

Finally it is important to stress that the best extension is one which cannot be easily detected by the average passery-by.

Domestic Extension Part 3

This deals with removing the existing external french door frame, making good and altering the opening size and extending the existing electrical system.

		£
a)	Carefully remove existing french door frame complete and cart away from site,	
	Labour: 2 men .25 day	25.11
	Plant: skip	6.00

b)	Re-form lower part of opening by removing brickwork under side lights and make good brickwork and plaster to reveals,	
	Labour: 2 men 1 day	100.46
	Plant: skip	10.00
	Material: say	15.00
c)	Extend existing electrical system in adjoining room to provide two light points and four power points,	
	By sub-contractor	350.00
d)	Allow for making good,	
	Labour: 1 man .50 day	25.11
	Material: say	10.00
e)	Allow for contingencies	300.00
f)	If the existing house has a hollow floor it will be necessary to allow for underfloor ventilation in the extension floor,	
	Allow provisional sum,	60.00
		901.68
g)	Allow for Preliminaries as before 12½%	112.71
		1014.39

Project 15
Rebuilding a chimney stack

For the sake of this project it will be assumed that the chimney stack involved is on a party wall of a two, or maybe three, storey Victorian house. Chimneys built in this era will almost certainly be in need of repair work or complete reconstruction by now as the brickwork or pointing may have deterioated, or the chimney may be leaning or in a dangerous condition due to the elements.

One of the many reasons for having to rebuild a chimney stack concerns domestic boiler flues which are unlined. Many houses built between the wars were provided with solid fuel domestic boilers, the chimney stacks being built in cement or cement lime mortar. During the post war years many houses changed to slow burning or gas fired boilers and this meant that the flue gases which contained quantities of sulphates condensed into liquids attacking the cement in the mortar and causing it to expand. As chimney stacks are frequently subjected to soakings from rainwater this dilutes the acid but in turn carries it to all parts of the brickwork. The prevailing wind will cause the brickwork to dry out at an unequal rate on the elevations and this causes the stack to lean.

A more common case is that of a new boiler connected to an old flue and causing heavy acidic condensation. This can normally be seen by salt crystalisation at the top of the stack and will not only affect the mortar but also the brickwork and the parging.

Scaffolding will be required in order to give a good and safe working platform around the chimney stack in question, although there are some specialist manufacturers nowadays who produce purpose made scaffold units for stack work.

In the example given above the chimney stack itself will be owned jointly between the two householders. It is unlikely that only one side of it will need to be rebuilt unless of course the chimney is in an exposed location where the prevailing wind has caused greater deterioration to one or two elevations of the chimney than the others. Even if only working on one side of the chimney a working platform around the whole stack may be needed and the written agreement of the adjoining owners will of course be required.

The chimney pots will be the first items to be removed and these may need to be rebedded once the rebuilding has taken place. Chimney pots are heavy and awkward items to manoeuvre and great care should be taken by the operatives on site not only to prevent the chimney pots being damaged but also to prevent anyone on site from being put in any danger due to them being mishandled.

When the old brickwork is being removed this, too, will have to be carefully lowered to the ground (and not just thrown from roof level as is a practice often carried out on jobs).

If the householder wishes all the new work to exactly match the old work, and the bricks themselves have not deteriorated too much, then these will probably need to be saved, cleaned up and reused again in the rebuilding stages. If the chimney stack is right in the middle of the roof as in this case then it may not be necessary to match exactly. But if the chimney stack is on the end elevation of, say, a pair of properties then the appearance will be all important.

Although it may be possible to reuse the existing bricks it should be noted that the existing bricks may not be the most suitable for the job. Most brick manufacturers will advise on the best type of brick for use in any particular situation and these should be in accordance with the current British Standard. There is a wide range of choice and there is a very good chance that a selection can be made nearly matching the original.

In theory any work which is carried out by a builder to the structure of a property requires Building Regulations Approval. It is unlikely, however, that the majority of Building Control Departments would be very interested in this type of work, but it is advisable that they are informed in good time that the work is to be carried out so that they may process an application should they consider it necessary.

Rather than leaving all the old debris lying around in the owner's front garden it should be cleared away from the site as soon as possible. A full size skip would probably not be required unless other works are being carried out at the same time and the builder may find it adviseable to hire one of the mini skips that are now widely available. The work may also be part of a roof covering renewal job and another way of safely removing debris to ground level and depositing it in a rubbish container skip is to purchase or hire a purpose made plastic chute. The debris is simply thrown into the chute which takes it safely and conveniently down into the skip.

Damp proof courses in chimney stacks are extremely important. Two dpcs should be inserted, one where the stack rises above roof level and the second below the very top of the stack, preferably a couple of courses from the top. It is important that the materials used in the damp proof course should form a continuous impervious membrane. Flexible materials such as bituminous felt or asphalt are preferable rather than a rigid material such as slate or engineering bricks.

Damp proof courses should extend through the full thickness of the stack including the pointing, rendering or other facing materials. The dpc itself should project from the face of the stack, or a metal flashing should be inserted immediately below it to avoid the possibility of rainwater running down the surface above and seeping in below the level of the dpc. Materials for damp proof courses are listed in the current British Standard and, as previously mentioned, a flexible sheet material with a sealed joint should be used. A bituminous dpc can be used as this should be impervious when it is given a 100 mm lap sealed with bitumen.

If the brickwork is spalling then the client may instruct you to simply rake out the joints to form a key and to apply a couple of coats of rendering. It should be understood, however, that this will not solve the problem as the bricks are likely to have been attacked throughout their thickness and the rendering will not hold. Obviously, if the stack is presently rendered or rough cast then it should be rebuilt and finished to match. It would be wise though to determine the cause of the problem prior to rebuilding and some investigative work will be needed.

Another favourite repair to try and save money is to reduce the height of the chimney stack, but once again if dampness is the problem then the remainder of the stack and the flashings, etc., would still be open to the elements. Also planning permission may be required before the appearance of the chimney stack can be altered, especially if the building is listed.

The rebuilding work itself should be relatively straightforward although great care should be taken to avoid disturbance of the occupiers of the house (and the adjoining property). The internal parging in the flue should be continued to the top of the new stack flush with the existing, taking care not to allow too much debris to enter the flue. Fig.30 details a typical chimney stack above roof level.

Needless to say, the flaunching, pots and cowls should be reinstated in accordance with proper practice and flashings renewed if necessary, preferably in Code 5 lead.

On completion the flue should be swept clean and all rubbish, plant and surplus materials removed from site.

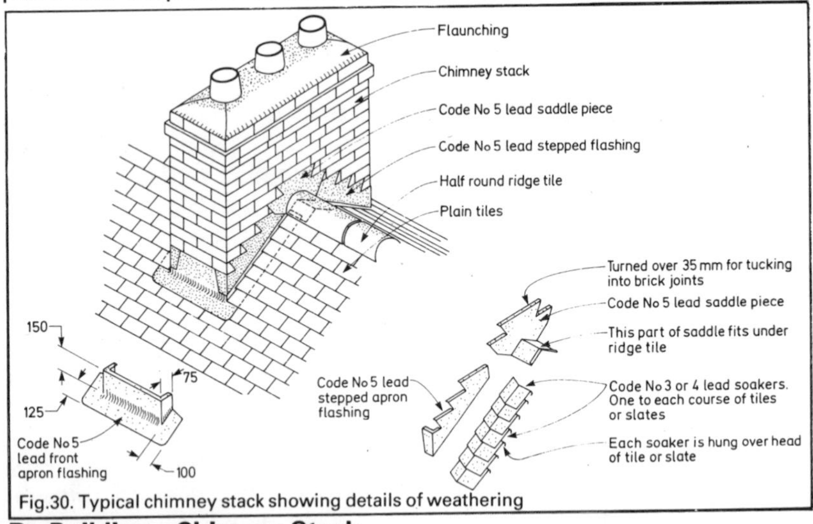

Fig.30. Typical chimney stack showing details of weathering

Re-Building a Chimney Stack

Take down and re-build stack as Fig.30

		£
a)	Allow to clear access for and supply and erect scaffold around stack with suitable platform for materials;	
	Labour: 2 men .75 day	75.35
	Plant: Scaffolding, say	100.00

b)	Carefully remove existing chimney pots and set aside for re-use if in sound condition;		
	Labour: 1man .07 days		3.52
c)	Carefully demolish existing brickwork to stack and set aside all bricks in good condition for future use after cleaning. Basket down all rubbish and including clear away old perished flashings and carefully remove and set aside ridge tiles either side of stack;		
	Labour: 3 men half day		75.35
	Plant: say		20.00
d)	Prepare brickwork to receive new; re-build stack using part old part new bricks (say 50-50) including parging flues, forming oversailing courses, setting pots in position and flaunching in cement and sand and provide and bed lead lined damp course at roof level and top of stack;		
	Labour: 3 men .75 day		113.02
	Material: Bricks (120)	24.00	
	Mortar	5.00	
	Dpc	7.50	36.50
e)	Rake out joints for and supply and fix all new flashings including new soakers, point joints, replace disturbed roof tiles and ridge tiles;		
	Labour:2.5 men one day		125.58
	Material: lead		60.00
f)	Remove scaffold, clear up all rubbish and remove from site and sweep flues;		
	Labour: 2 men one day		100.46
	Plant: skip		20.00
			729.78
g)	Allow for prelims 12½%		91.22
			821.00

Various reductions to the foregoing rates could be achieved depending upon the condition of the existing bricks and flashings but it must also be borne in mind that the adjoining property may well be affected by this work and any likely costs arising therefrom must be taken into account.

The foregoing labour rates are based on £4.60 per hour for a 39 hour week plus 40% for profit and overheads. Adjustments should be made accordingly for any variation to these.

Project 16
Removing the whole chimney breast and the stack

It is obviously important to determine the client's reasons for wanting a chimney breast removed. If he requires the ground floor breast to be removed completely from a two storey house, and the chimney stack at roof level is in a poor condition then it will probably be wise to remove the whole chimney from the ground to the roof completely. This may, however, involve some unnecessary work and might indeed even work out costlier than just carrying out the original required job. Of course, greater room would be obtained in one of the first floor bedrooms above, but nevertheless the whole job should be given consideration and thought.

Many properties have just had the ground floor chimney breast removed with absolutely no support, or even thought for the first floor chimney breast above. This leaves all the brickwork to the first floor chimney breast in an extremely unstable condition with even the possibility of a complete collapse.

In order to remove the whole chimney breast and chimney stack Building Regulations Permission (or approval under the London Building Acts) will be required, and drawings may need to be produced depending on the local authority involved. It is wise to make preliminary inquiries to the Building Control Department and to adhere to its wishes.

If the chimney breast is on a party wall then there is a high possibility that the chimney stack will be shared between the two houses. If this is the case then a Party Wall Agreement might be

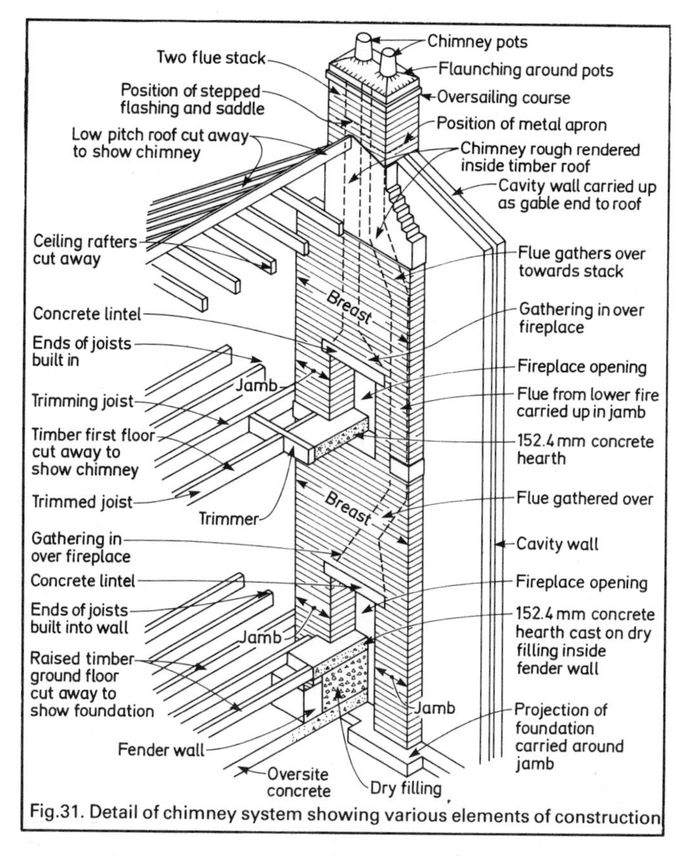

Fig.31. Detail of chimney system showing various elements of construction

necessary, or in the very least some form of written approval from the adjoining owner. If 50 per cent of the width of the chimney stack is to be removed then it may become unstable and you could find that total removal is not possible.

A structural engineer might be necessary to ascertain the structural stability of a chimney stack of only half its original width.

If the whole chimney stack is to be removed then many points raised in Project 15 will also apply in this case. A scaffold will need to be erected to provide a working platform, and care will need to be taken in removing the pots and old bricks, etc.

Take the stack down to below roof level, and if it is not possible to complete the making good of the roof in the same day then ensure that adequate temporary protection is provided against the elements. It is advisable that once the stack is removed down to below roof level, rather than proceeding with this removal it is best to make good the roof immediately. If there is a party parapet wall above roof level and an adjoining stack then this may need to be made good by cutting out various areas of the brickwork and repointing the whole elevation or applying a couple of coats of rendering. The rafters will almost certainly be trimmed around the existing chimney stack, and rather than remove the trimmers and extend the rafters it is far better to simply insert new timber members, extend the ridge board and ensure that there are adequate fixings for the extended roof coverings.

Flashings, etc., will also need to be extended to ensure adequate protection against the ingress of rainwater. Other areas around the chimney stack and party parapet wall will also need to be made good such as flaunchings and parapet wall copings.

Removal of the chimney breast from this stage should be made from inside the roof space working downwards towards ground floor level. Be careful to protect the occupants' belongings and furniture, etc., as you will probably need to have the run of the majority of the house. A rubbish skip will be essential to remove all

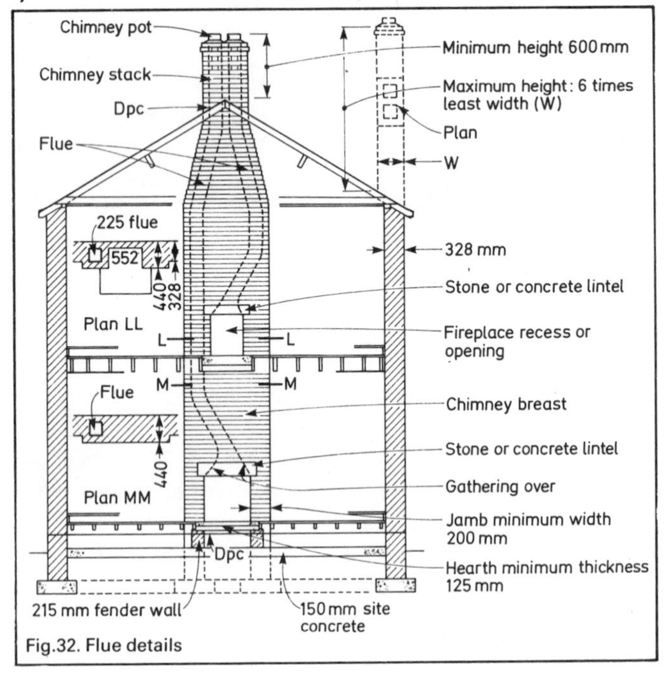

Fig.32. Flue details

the old brickwork, etc., and you will probably need to spend quite a large amount of money on the hiring of dust sheets! Sometimes it is almost as important to the occupier of a house to know that his furniture, etc., has been treated with care and respect as it is for him to know that the job has been carried out to a high standard. This is never more so than during a job such as this and it is important to realise (and to allow for in the estimate) that working in occupied premises is likely to add quite considerably to the cost of the job.

When the chimney breast is being removed completely it is recommended that any hearths are also removed. Not only are they unnecessary, but the floor joists will need to be retrimmed and altered to allow for the fact that the floor areas have been increased.

If the trimmers are to be removed at first floor level then it will be necessary to support the floor from below, preferably with Acrow props to properly spread the load over the floor joists below. This project will not endeavour to detail all the structural points regarding this aspect of the job, but suffice to say that chimney breasts are as much a part of the structure of a house as the walls and floors themselves, and it should not be taken for granted that everything else will stand and be supported once they are gone.

Another point to consider when estimating for the job is that as the chimney breast has to be removed at high level in a room then

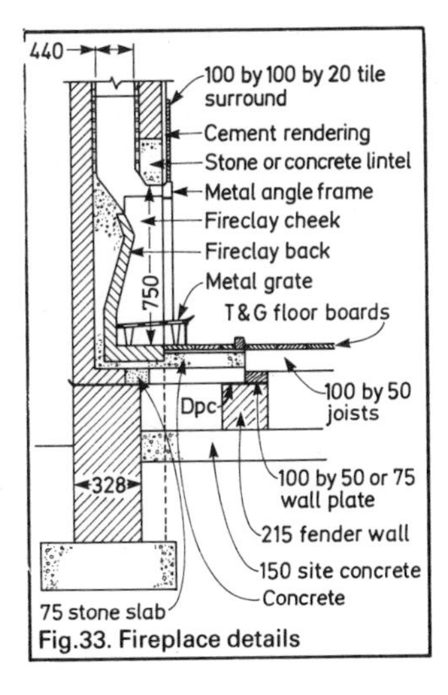

Fig.33. Fireplace details

a proper working platform will have to be provided to carry out the work. This will also apply when the making good is carried out to the walls and ceilings, etc.

When an item such as a chimney breast is removed it is extremely difficult to disguise the fact that it was once there unless the subsequent making good is carried out to an excellent standard. No matter how expert the plasterer may be it is often difficult for him to provide a smooth and even surface if he starts off with an extremely poor and uneven background.

Obviously there will need to be making good to the brickwork and wall plaster, and thorough preparation and cutting back of the surrounding plastered areas is the only way of disguising the previous use of the wall. If the surrounding wall plaster is defective anyway it would probably be a good idea to allow for the hacking off and complete renewal of all the plaster to that one wall.

Ceilings, floorboards, skirtings, plaster and brickwork, etc., will all have to be made good - it may be necessary to completely renew some areas such as plaster as already mentioned, skirtings and a fair proportion of the ceiling. It should be noted however, by the client whether or not you have allowed for complete renewal of wall plaster, etc., or only partial renewal because if you are in competition with other building contractors this will certainly make a difference to your competitiveness.

Finally if only the lower part of the chimney breast is to be removed then the structure above should be properly supported by either corbelling the brickwork or providing suitable lintels, etc. Remember that redundant chimney flues should be thoroughly ventilated to alleviate the possibility of gases building up, condensation forming and the dampness not being able to thoroughly dry out.

Removal of chimney breast

For this example we have assumed that no adjoining owner is involved and the stack will be demolished completely (as Fig.31)

		£
a)	Clear access for and supply and erect suitable scaffold for removal of stack;	
	Labour: 2 men .75 day	75.35
	Plant: Scaffold	100.00

b) Remove pots and flaunching and demolish stack to below roof level carefully lowering all materials to ground level;

Labour:3 men .33 day		50.23
Plant: say		10.00

c) Make good opening in roof including supplying and fixing all new timbers for rafters, ridge etc; fix felt and battens and tile to match existing;

Labour: 3 men one day		150.69
Materials: Timber	7.00	
Tiles etc. Felt	30.00	30.00

NOTE: If it is not possible to demolish the stack and make good the roof in one day allowance must be made for overnight weather protection.

d) Demolish chimney breast in roof space and basket out materials via loft access and including providing a safe working platform and protection to the ceiling under. It has been assumed that adequate loft access is available;

Labour: 3 men half day	75.35
Plant: skip	10.00

e) Carefully demolish chimney breast at first floor level including removing hearth, fireplace etc. and leave ready for new plaster work (approx. 5.00 SM);

Labour: 3 men one day	150.69
Plant: skip	16.00

f) Carefully demolish chimney breast at ground floor level including removing hearth, fireplace etc. and leave ready for new plaster (approx. 6.00 SM);

Labour: 3 men one day	150.69
Plant: skip	20.00

g) Prepare for supply and fix necessary timber joists and bearers to infill openings in ground and first floors and first floor ceiling. At ground and first floor levels, cover joists with boarding to match existing;

Labour: 2 men .75 day	75.35
Material:	100.00

h) Make good ceilings at ground and first floor levels with single or double layer plasterboard as necessary and set in plaster to match existing;

Labour: 2 men half day	50.23
Material:	15.00

i) Dub out in cement and sand and render and set walls to match and align with existing (approx. 11 SM);

Labour: 2 men 1.50 days	150.69
Material:	55.00

j) Allow to replace picture rails, skirtings, dado rails etc., all to match existing;

Allow Provisional sum	100.00

k)	Allow for protection of furniture and floor coverings and cleaning up at the end of each working day; Labour: 1 man one day Plant: say	 50.23 30.00
		1465.50
l)	Allow for prelims 12½%	183.19
		1648.69

NOTE:

1 No allowance has been made for any decoration as it is assumed a major re-decoration would be involved in addition to the work we have dealt with.

2 In older properties the condition of the existing plasterwork and possibly brickwork may be such that areas additional to those directly concerned with removing the chimney breast will also have to be dealt with. A close inspection before the price is submitted is essential and the appropriate adjustments made or a clear understanding of the extent of your work is agreed with the client.

3 The price must also reflect the needs of the occupants of the premises in terms of access and working hours.

The foregoing labour rates are based on £4.60 per hour for a 39 hour week plus 40% for profit and overheads; adjustments should be made accordingly for any variation to these.

Project 17
Damp proofing and timber treatment

Insertion of a chemical damp proof course does not normally cause the disruption that timber treatment would, but for the sake of this project it has been assumed that the property is unoccupied and without fitted floor coverings, etc.

Although the damp proofing works will be carried out by a specialist contractor, it is important for the builder providing attendance to know the reasons for requiring the damp proof course.

In a lot of Victorian terraced houses, common around South London and North Surrey, one layer slate damp proof courses were originally built into the brickwork. These were inadequate not only because they do not provide a monolithic and impervious layer but also they tend to break down and crumble. Dampness is then able to rise between the joints or through the deteriorated slate itself.

This is a plain and simple case for a chemically injected damp proof course but there may also be other reasons why dampness is entering the property. In many instances there has been a build-up of paving around the property which will have bridged the damp proof course. Although it is not always necessary to lower the level of the external paving when a chemical damp proof course is injected, in the majority of cases it is best to do so anyway.

If it is not possible to lower the paving then that part of the paving adjacent to the property can be dug up and the level reduced and a few inches of pea shingle placed in the bottom of this newly formed trench to reduce the overall level and allow rainwater to escape with ease as in Fig.34.

When the wall is having a damp proof course injected it will always be necessary to hack off the plaster internally up to a height of approximately one metre. This also means, of course, that skirtings and probably some architraves will need to be removed and carefully laid aside for possible re-use at the end of the job. If the dampness has been allowed to remain for any length of time, however, it will be necessary to renew all the timbering such as skirtings, etc., with previously treated joinery to match.

The specialist contractor can now attend site to drill holes in the external (and internal walls if necessary) and insert his chemical solutions under pressure.

While this work is in progress you can be lifting floor boards throughout the property at the rate of approximately one in every five in preparation for the timber treatment. If the property does happen to be occupied, moving furniture and lifting floor coverings should be carried out by the occupier or by a specialist carpet fitter. Remember that when lifting the floor boards in readiness for timber treatment you may have to remove a number of the skirtings first.

Access to timbers in the staircase may not be possible as a permanent soffit may be fixed to the underside. In this case it will be necessary to drill holes in the timber risers of the staircase and the size and spacing of these holes will need to be directed by the timber treatment specialist.

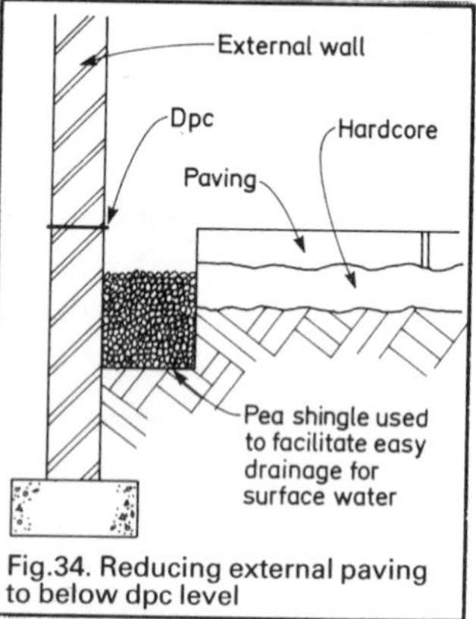

Fig.34. Reducing external paving to below dpc level

Access into the roof space may not be possible as there are still many older houses that do not have an access hatch at first floor level. Allowance should be made for this and, if the client wishes it, a permanent roof space access hatch should be made and fitted, otherwise you will have to cut a hole to gain access and make good the hole on completion of the job.

Roof spaces are generally very dirty places and the timber treatment specialist will have to clear out any rubbish, lift any roof insulation and clean off all the ceiling joists and rafters, etc. As the main contractor you may be required to carry out these items for the specialist and it should be realised by the client that it is not always possible to re-use the existing roof insulation, especially if it is old and has become compressed.

While in the roof space you should inspect the ends of joists and other areas of timbering for wet rot. The roof might have been leaking over a period of years causing such problems, or the guttering joints could also have caused wet rot to the fascia and soffit boards. Remember that what appears as wet rot on an exposed timber may give rise to suspicions that the same conditions are producing dry rot in timbers hidden from view and, as such, unventilated. Attacks of woodworm can also spread to hidden timbers and continual infestation in relatively confined areas cannot be ignored if the conditions are favourable.

In this country there are four types of wood boring insect which can cause structural damage to timbers. These are: the Common Furniture Beetle (*Anobium Punctatum),* the House Longhorn Beetle (*Hylotrupes Bajulus),* the Death Watch Beetle (*Xestobium Ruforillosum)* and the Lyctus Powder Post Beetle (various species). Where there is no actual structural failure or collapse the only sign of these wood boring insects is in the form of the exit hole produced when the fully grown beetle emerges from the wood. Only the previously mentioned beetles require treatment and a specialist knowledge is required to determine the type of beetle present in any section of infested timber.

Once the timbers to be treated have been sprayed, the floor boards can then be replaced. The timber treatment specialist will also need to spray the tops of the boards themselves and any other untreated timbers. The rooms should then be well ventilated and preferably not occupied for a few days afterwards. It can be seen that occupation of the house during the course of such works is not advisable.

British Standard 4072:1966 deals with wood preservation. If necessary, a list of the manufacturers of liquid insecticides and of suppliers of preservative treated timber can be obtained from the British Wood Preserving Association, 62 Oxford Street, London W1. It also provides guides to the safety precautions to be taken when the chemical treatment is carried out.

As previously mentioned, the dampness on external walls could well have given rise to wet rot in the adjoining joists and wall-plates, etc. Some timber treatment companies insist on carrying out their own cutting out and replacement of affected timbers although you may well be asked to carry out this work under their supervision.

If a suspended timber ground floor is not supported on brick sleeper walls, it might be necessary to renew the whole joist if it has been affected. Other than this it is best to temporarily support the floor, cut out the affected timbers back to at least 150 mm beyond the rot and renew as existing, ensuring that all timbers are properly pretreated. Also be sure to relay a damp proof course below any wallplates.

Sub-floor ventilation in older houses, and indeed in some new houses, is often inadequate. Sometimes these houses are extended with a solid floor extension taking no account of the sub-floor ventilation to the existing timber suspended floor of the original property. Another reason for lack of ventilation is blocked and choked vents, and sometimes even paving that rises above the vent itself.

Dry rot (*Merulius Lacrymans)* needs two conditions in which to grow: lack of ventilation and dampness. Even if an area below a timber suspended ground floor has some dampness this will quickly dry out if the ventilation around the structural timbers is adequate. This is not to say that rot will not occur in this situation but adequate sub-floor ventilation will definitely reduce this possibility.

As previously stated, the reasons for the occurrence of dampness should be ascertained and the causes rectified. Defective pointing should be hacked out and renewed, as should defective rendering. If the bricks themselves have become porous then apart from rendering the entire property, which could look aesthetically incorrect, it may be possible to apply a coat of silicone water repellent. 'Synthasil' is an ideal material as not only does it reduce the porosity of the bricks but, being a clear coating, it does not affect the look of the building.

The drill holes made by the damp proofing specialist will need to be made good and any rendered plinths at low level will need replacing. Care must be taken at this point not to bridge the damp proof course, as this can result in further outbreaks of rising damp.

Internal replastering should be carried out by either applying a render coat (making sure to include a salt retarder), and a float and set coat of 'Sirapite' plaster. Make sure that no 'Carlite' plasters are used as these tend to retain the water and nullify the specialist damp proofing guarantee.

A more recent innovation is the use of 'Tilcon Limelite' plasters. These are basically a renovating plaster and while not cheap are ideal for use in this situation.

The property may also be suffering from dampness to solid kitchen floors which which have never had the benefit of damp proof membranes. These floors will need to be broken up and removed and a new floor laid, maybe even at a different level, incorporating a damp proof membrane. Be sure that this membrane ties in properly to the external dpcs and that adequate laps are made.

Attendance on damp proofing and timber treatment

There are many different specifications in this type of work, involving varying degrees of attendance. Normally the specification provided by the specialist sub-contractor sets out clearly what he requires of the builder. Only guidance on the type of work and its requirements can therefore be given.

1. Damp Courses

Where these involve taking down plaster a considerable volume of dirt and rubbish is involved, therefore care must be taken in occupied premises to cover up and protect and damp down. Skirtings, architraves, dado rails, etc., must be removed together with any timber grounds and where necessary eventually replaced with similar pattern new material. As some old properties contain unusual moulding or large section architraves and skirtings it may be necessary to have these specially made at resultant high cost. It may also be necessary to remove doors and frames.

Before attempting to price work of this nature it is necessary to clearly determine and agree the full extent of your responsibility and to commit this to writing.

The re-plastering of walls may also be carried out using a waterproof mix: in older properties it will probably be necessary to dub out, and due allowance must be made for this.

Due to the very uncertain nature of much of this attendance work it is frequently carried out on a time and material basis (daywork) and reference to the Guide to Estimating Building Work published by BTJ will give guidance on calculating suitable hourly rates and will also give guidance on the unit prices for the measured work involved. We have not quoted any prices here as they will vary enormously depending upon the quanities involved, access, working conditions, etc.

2. Timber treatment

The text has already covered the problems involved, particularly in the case of occupied properties.

It is suggested that a check list of items likely to arise is prepared before the job is priced. Each item on the list can then be dealt with on its merits. A typical list could consist of the following:

1. Lift floor boards
   - i) floor covering
   - ii) furniture
   - iii) skirtings
2. Clear out floor void
3. Replace boards
4. Roof access
   - i)form new access
   - ii) clear out loft space
   - iii)remove insulation
5. Supply new floor boards
6. Supply new roof hatch
7. Supply new insulation
8. Etc.

Project 18
Replacement of flat roof coverings

The roof in question is assumed to be over a small office or industrial unit and is probably only part of the whole building. Therefore as the premises are commercial there are various aspects to be taken into consideration that would not necessarily always matter if the job was over a domestic property.

Access will probably be reasonably straightforward as the unit will almost certainly have good access for its own day to day business. However if the roof is at the rear of the building it may be more difficult to get to. You may find that the roof backs onto a railway or similar and careful consideration should be given to the difficulties in moving materials and plant, etc.

It is also assumed that the roof is over a single storey building, but you may find it prudent to erect a small tower with a hoist of some description if the need arises.

As in many building projects a rubbish skip will almost certainly be necessary unless the roof is so small that all the debris and rubbish, etc., can be loaded into the back of a pick-up truck. A rubbish skip and builder's van, etc., will result in the loss of spaces if there is a car park to the commercial unit and the clients should be made aware of this.

The existing roof coverings are likely to be of the traditional built-up three layer felt type with various upstands to parapets, etc. It is common practice nowadays to substitute for the three layer felt a higher quality polyester based material. Various manufacturers of roofing products and components provide comprehensive trade literature and technical advice. Contact one of the larger

This flat roof with cracks, ponding and flourishing plant life is in urgent need of repair.

manufacturers through their advertisements in magazines such as Building Trades Journal and seek their advice on any specific job.

Replacements of asphalt roofs and flat metal sheet roof coverings have not been mentioned as these are almost always carried out by specialists and it is not usual for a small general builder to undertake such jobs.

You should determine the type of construction of the flat roof and advise the client that if the existing roof coverings have deteriorated really badly it is likely that the roof decking underneath and maybe even the roof structure itself will be in need of some repairs. It is obviously not always possible to know what these problems will be at the time of estimating but nevertheless the client should be made aware that such repairs may need to be carried out.

The materials as previously mentioned will vary very little with regard to their weight and so I doubt if additional strengthening of the flat roof will be necessary. If a different type of roof decking is used then any additional weight will of course result in the necessity to strengthen the roof structure.

You may find that the existing roof covering has reached the end of its useful life prematurely due to the unsuitability of the materials in that particular location. Once again expert technical advice from the various large manufacturers is not only helpful but most of the time it is free (although they would of course like you to buy their product). The current British Standard deals with the roof coverings and the correct ways in which they should be used. It would pay dividends to buy a copy and study it carefully.

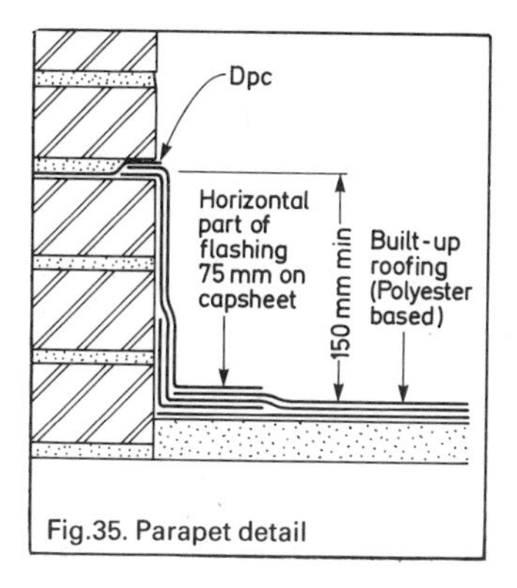

Fig.35. Parapet detail

Many existing small flat roofs have been covered with a layer of stone chippings to reflect the extremes of sunlight and temperature and to generally provide protection for the roofing felt. If the client insists on a similar type of roof covering being used then it may be necessary to carefully remove the stone chippings and set them aside for reuse. One of the problems encountered over the years with these stone chippings, however, is that the felt underneath the chippings can sometimes be easily damaged by people walking over the roof and treading the stones into the felt.

It is unlikely that temporary weatherproofing will be required over a small roof in a straightforward situation but nevertheless the possibility of its use will need to be considered. The supply and erection of temporary weatherproofing will of course add greatly to the overall cost of the job.

One of the reasons why the roof may have deteriorated in the first place may be due to the lack of proper falls. This would result in rainwater ponding on the roof causing the breakdown of the roof coverings and consequently shortening their life. Due consideration to this problem should be given and a section of the estimate set aside for the additional costs that this would incur. The only real way to overcome this problem would be to remove the roof decking and, assuming that the structure of the roof is timber joists, adding firring pieces to the tops of the joists prior to replacing the roof decking. As previously mentioned the roof decking and structure should always be inspected not only to ensure that they are still capable of carrying out the job for which they were intended but also to ensure that the roof meets with present day requirements for ventilation and insulation.

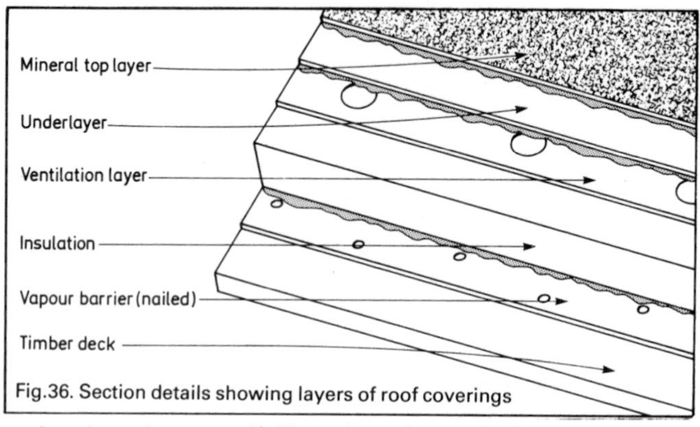

Fig.36. Section details showing layers of roof coverings

There is also the possibility that the existing insulation to the unit may be asbestos based. Current regulations prohibit the use of asbestos insulation in this way but as long as the insulation is properly sealed and has not become damaged it is unlikely to cause any harm to the occupants of the building. However, should the client wish this form of insulation to be removed it is a very delicate operation and specialist licensed companies must be used to tackle such a job.

If additional insulation is required or if the present insulation is resulting in excessive heating costs then it is recommended that a Jablite insulation or similar is used. It manufactures Thermodeck which in my experience is both economical to buy and easy to install. Jablite also provides a good range of technical literature with its products.

Ventilation to the voids in the roof is also extremely important especially if the deterioration of the roof and roof deck is due to lack of such ventilation. The current Building Regulations state that roof voids should be adequately ventilated and there are now some proprietary types of ventilator on the market. Different types of ventilators will obviously need to be used in different situations.

The access to the site was mentioned previously and this also brings us to the storage of materials on site. Precautions should be taken to ensure that the materials do not block any vital passageways or that they are not in such a position as to invite theft or vandalism. Be sure to consult with your client about such aspects of the job when obtaining details for preparation of the stimate.

If you are working under a minor works contract then the job may be subject to a penalty clause if there is a time delay. Be sure that you have adequate labour and time to carry out the job and remember that inclement weather will greatly affect roofing works in general.

Needless to say all works of this description and the laying of roof coverings should be carried out strictly in accordance with the manufacturers' instructions and specifications and also in accordance with the various British Standards and Codes of Practice. As previously mentioned the roof coverings should be laid to the correct falls and obviously all the proper laps of the sheet roof covering should be allowed for in estimating for the amounts of materials as should the upstands to the parapets and chasing into brickwork, etc.

If the small builder is in any doubt as to his capability of laying flat roof coverings, even if he does have a comprehensive manufacturer's guide then the works should be left (or sub-contracted) to specialists.

If the defective roof coverings are not the result of ordinary wear and tear then the client might have a case for claiming under his building insurance. Such a claim should also include for damage to any internal areas such as suspended ceilings, etc. You may also need to estimate for repairs to the ceilings and internal areas or to get advice from (or once again sub-contract to) suspended ceiling specialists.

Lastly you may find there are various pipes, flues, etc., protruding from the roof itself. As with the parapets and upstands it is sometimes better to use metal flashings (some manufacturers provide preformed sections for this purpose) rather than try to form the flashing in the same material as the roof covering.

Replacement of flat roof coverings

For the purpose of this project we have assumed a single storey building with three external walls with parapets, approximately 10.00×4.00×3.40 metres high. Easy access; a working space is available.

		£
a)	Supply and erect scaffold tower with Gin wheel and fall rope;	
	Labour: 2 men .25 day	25.11
	Plant hire: including transport on and off site	80.00
b)	Carefully remove existing roof coverings including cutting out pointing where necessary and lower to ground for carting away (approx. 62 SM)	
	Labour: 2 men .50 day	50.23
	Plant: skip	20.00

c)	Allow Provisional Sum for repair and/or replacement of defective areas of roof decking,	250.00
d)	Clean and prepare for and supply and lay three layer high performance built up roof covering as Permanite "Superflex" System (warm roof construction) including provision of new insulation;	
	Labour: 2 men 2.50 days	251.15
	Material: Bituglass type 3B - 40 SM	45.00
	Coolag roof boards (50mm) - 40 SM	335.00
	Bituglass underlay type 3G - 62 SM	73.00
	Superbase underlay - 62 SM	123.00
	Superflex AA - 62 SM	205.00
	Sundry materials, mastic etc.	30.00
	Plant:	25.00
e)	Repoint flashings, clear down surplus materials and load for cart away;	
	Labour: 1 man .75 day	37.67
	Material:	5.00
	Plant:	5.00
f)	Take down scaffold and clear site;	
	Labour: 2 men .50day	50.23
		1610.39
	Allow for prelims 12½%	201.30
		1811.69

The foregoing labour rates are based on £4.60 per hour for a 39 hour five day week plus 40% for profit and overheads. Should any of these factors differ from those to be used, appropriate adjustments should be made accordingly.

Project 19
Renewing pitched roof covering on an end terrace house

The ease with which houseowners in the past couple of years have been able to obtain local authority repair grant aid has meant that many owners of turn of the century properties have been able to renew their roof coverings. It is obviously important, therefore, that builders know how to accurately estimate for such work.

There are, by necessity, many similarities between this project and project 18 concerning the replacement of flat roof coverings. For the sake of this project it has been assumed that the job in question is to be carried out on a turn of the century end of terrace house with a slate roof covering and cast iron gutters.

It is a legal requirement these days that some form of scaffolding is erected when this type of work is being carried out, and the scaffold must comply with the relevant construction regulations regarding working places.

Once again, access to the site and to the roof in particular will be an important factor governing the moving of materials and debris. It is possible to hire a chute which will move safely old roof coverings into rubbish container skips. Some scaffolding contractors provide and fix these as an integral part of the scaffold.

While talking about scaffolding it should be mentioned that if comprehensive structural repairs are being carried out to the roof timbers or if a room is being constructed in the roof space itself then a temporary covering may need to be provided over the whole house. This would of course consist of a system of scaffolding with waterproof material to keep out the weather. This will not normally be needed where a straightforward roof covering renewal is being undertaken.

Unfortunately many householders nowadays renew their traditional slate covered roofs with concrete interlocking tiles. Although planning permission will not normally be required except where a listed building is concerned, the replacement of a smooth grey slate roof covering with a sand finished pink coloured interlocking roofing tile can quite often look aesthetically incorrect. Therefore the look of the building should be taken into consideration although the majority of householders will probably pay more attention to the overall cost!

New or even secondhand slates are prohibitively expensive so many roofing specialists are recommending the use of what used to be known as asbestos slates. Eternit manufactures roof tiles called Series 2000. These can sometimes look very attractive in place of slates and the cost is quite competitive with that of the cheapest form of concrete interlocking tile.

Concrete interlocking tiles are fairly heavy and the roof structure may well need to be strengthened to allow for this additional weight. In this case Building Regulations approval (or consent under the London Building Acts) will need to be obtained. In fact most local authorities now require an application to be submitted anyway to show details of the roof structure and the new roof coverings in comparison with the old.

Many householders store belongings in their roof space and these should preferably be moved or, at the very least protected while the roofing works are in progress. This is also true of cold water storage tanks, central heating expansion tanks and any pipes, etc.

Once the old slates have been removed (these roofs quite often deteriorate due to the failure of the nails rather than the slates themselves) the battens too will need to be removed and any old nails pulled from the rafters prior to any new work commencing. It is assumed also that the old cast iron gutters are being renewed and so therefore all of the roof timbers such as rafters, ridge board, fascia, soffits, etc., should be thoroughly inspected and any repairs or replacements carried out. My own thoughts are that it would be wise at this stage to treat the tops of rafters, etc., with a preservative against rot and beetle attack.

Strengthening of the roof structure will need to be inspected by the building control officer or district surveyor and calculations for this work may be required. Ensure that the loadings are evenly distributed onto the loadbearing walls and that any struts are properly birdsmouthed where necessary or supported as approved.

Untearable roofing felt should be used to cover the rafters and pretreated battens properly fixed to the correct gauge with galvanised nails.

Depending on the tiles used, the pitch will dictate the necessity of the amount of fixing nails required and if the pitch is extremely steep every tile may need to be twice nailed and clipped. The various valleys are also likely to be in a poor state of repair and these should be replaced with Code 4 lead. Some manufacturers make purpose-built valley tiles but these are normally few and far between and only used where plain tiles are involved. They can look attractive but usually require special timber boarding to be fixed to the valley on the underside of the tiles. Should lead prove to be prohibitively expensive the use of zinc or even Zincon may be preferable.

As mentioned in Project 18, ventilation to roof spaces is seen to be important nowadays. Most roof tile manufacturers have various forms of ventilator tile on the market, the most popular of these probably being the ridge tile ventilator. A decent through flow of ventilation is necessary and either airbricks should be installed in any gable end walls or some form of fine mesh (once again there are many proprietory brands on the market) be fitted at regular intervals in the soffits at eaves level. This is especially important once the roof has been felted and insulated. *See Fig.37.*

When estimating for renewal of roof coverings it is important to inform the client of any other works that would be needed at roof level. On this type of property it is highly likely that repairs or

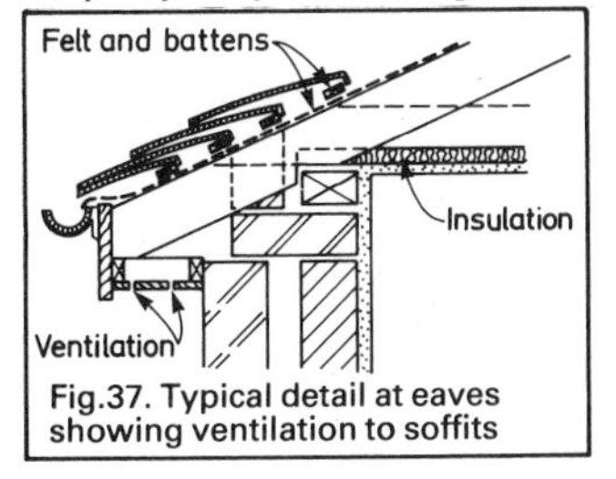

Fig.37. Typical detail at eaves showing ventilation to soffits

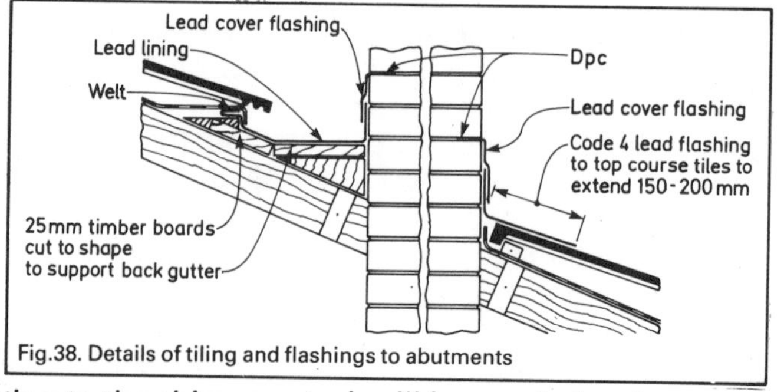

Fig.38. Details of tiling and flashings to abutments

repointing to the chimney stack will be needed and renewal of old cast iron guttering. Where grant aid is concerned all this type of work will be required by the local authority to put the house in general and the roof in particular into a good weathertight condition. It is advised that guttering and downpipes are replaced with the more modern black plastic type. There are many brands available and probably the most important part of the job is to ensure that the gutters are fixed to the correct falls. Guttering will not be discussed in great detail as storm water drainage in general will be dealt with in a future project.

The junction of many chimney stacks to roofs is often finished with a cement fillet. This is not a very satisfactory method of ensuring that rainwater does not penetrate beneath the roof tiles and a Code 4 lead flashing should be used over the soakers. Although there are many preformed sections on the market it is probably best that an experienced leadworker is employed for this part of the job.

Upstands to party parapet walls including any rendering or repointing in these areas should also be estimated for and, if party parapet walls are rendered or indeed if any areas of flaunching are renewed or repaired, then a coat of Unibond should be applied to the brickwork first.

Many people are not too concerned as to what the end product looks like as long as the price is right. It is important, however, to try and achieve something that will be in keeping with the style of the property and the area in which it is located. There are many special tiles on the market which will greatly enhance the look of the job such as crested ridge tiles, finials and hip end tiles. Leadwork to flashings, etc., will of course greatly improve the look of the roof and also reduce the amount of maintenance necessary.

It is also a good idea to use a colouring in the mortar when flank wall end tiles are pointed up to try and tone down the harshness of ordinary sand and cement. This should also be used when pointing ridge tiles and parapet wall tiles and is definitely worth the small amount of additional time and expense involved.

Finally, the client must be made aware of the advantages of insulating the roof space should this not have already been carried out. This work does not of course need to be undertaken when the roof coverings are being replaced as in the previous project, but Government grants are still available for helping to finance the costs of domestic roof insulation and builders should inform clients of the economic value of providing insulation.

Replacement of pitched roof covering

For the purpose of this project we have assumed a two-storey end terrace house with hipped end, 6.00 metres wide×7.50 metres deep with a 40° pitch and one chimney stack on the party wall.

			£
a)	Supply, erect and dismantle scaffold to front, flank and rear of property (135 SM);		
	Labour: 2 men three days		301.38
	Plant:		100.00
b)	Remove contents of loft or protect and cover up and protect water tanks, pipes etc.;		
	Labour: 2 men .50 day		50.23
c)	Strip existing slates, battens, hips and ridges and lower to ground and cart away (approx. 70 SM) and remove nails from rafters;		
	Labour: 2 men two days		200.92
	Plant: skip		56.00
d)	Remove existing cast iron gutter and rainwater pipes and cart away and prepare for new gutters and pipes;		
	Labour: 1 man one day		50.23
	Plant: skip		6.00
e)	Supply and lay new reinforced roofing felt, battens and interlocking concrete tiles to all roof slopes;		
	Labour: 2 men 2.75 days		276.27
	Material: felt	50.00	
	battens	35.00	
	tiles	260.00	
	sundries	20.00	365.00
f)	Supply and fix new ridges hip tiles and bed in cement and sand including necessary hip hooks;		
	Labour:2 men .50 day		50.23
	Material:		54.00

g)	Supply and fix stepped flashing including all necessary raking out of joints and pointing and supplying and fixing lead slates;	
	Labour: 1 man 1.50 days	75.35
	Material:	120.00
h)	Supply and fix 112mm UPVC gutters and downpipes including all fixings and fittings;	
	Labour: 2 men 1.25 days	125.58
	Material:	102.00
i)	Allow a Provisional Sum to repair or replace defective fascia, soffit or rafters;	250.00
		2183.19
j)	Allow for prelims 12½%	272.90
		2456.09

NOTE: The foregoing labour rates are based on £4.60 per hour for a 39 hour five day week plus 40% for profit and overheads. Should any of these factors differ from those to be used appropriate adjustments should be made accordingly.

Project 20
Works to small bay window - I (replacement of bressummer beam)

A bressummer is normally a large timber beam that bridges the opening formed by the addition of a bay window in a house. Back at the turn of the century, when these were a popular feature of small dwellings, the beams were constructed of large sections of timber. Frequently nowadays, however, these beams are in need of replacement as they have either been attacked by dry rot, wood boring insects or wet rot, owing to their lack of protection from the elements.

When replacing a bressummer it is important to consider the material for the new beam. Although timber is still acceptable for structural use (it is still in fact one of the best materials available), the cost of such a large section of timber is not always an economic feasibility.

It is far more common nowadays to use either a rolled steel joist, a universal beam, a reinforced concrete or pre-stressed concrete lintel, or a lightweight steel lintel such as Catnic or similar. The new beam used will depend on several factors: the span and loadings involved, the wishes of the local authority and, of course, the cost.

This is the first of three projects concerning bay windows; the second and third will discuss underpinning of the bay and tying a bay into the main building where differential movement has occurred. It is quite possible though that all three of these repairs will need to be carried out at the same time, because they can all be caused by the same initial problem. Differential movement of the bay in comparison to the main building would probably be due to the base of the bay being constructed off a much shallower foundation.

It has been assumed that it is only the bressummer beam that is being replaced, and therefore any other work will be dealt with in Projects 21 and 22.

The householder should vacate the front reception room of his home for the duration of the job and ideally a dust screen should be erected halfway down the hallway to separate the front of the hall and the front reception room from the rest of the property. To remove the bressummer it will be necessary to support the various loadings imposed upon it, and in this case you should refer back to Project 3 regarding the formation of a through room. The first floor joists above may bear onto the front wall of the property and these would of course need to be properly supported. Scaffold boards and Acrow props should be positioned to effect this support but first the joist ends of both the ground and first floor levels should be inspected to ensure that the joist has not rotted away and that the loads can be properly transposed from one floor level to another.

While discussing rot in other structural members it should also be mentioned that any lean-to rafters or flat roof joists to the bay window that abut the main wall could also be rotten. These too should be exposed and inspected and any further supporting carried out.

Usually, a brick load above an opening will not involve all of the brickwork vertically above that opening right up to the roof level. There is normally a 45 deg. angle of collapse (*see Fig.39)* although you should still pin through to provide support.

If any slates or tiles are removed from the lean-to over the bay these should be removed for re-use and timbers inspected and repaired/replaced as necessary. All new timbers used and the timbers exposed should be thoroughly treated. If, however, dry rot is the cause of the problems all of these timbers should be cut out and any plaster hacked off back beyond the problem. Other precautions also need to be taken in the case of dry rot (*see Project 17*).

Internal plastering and external rendering will almost certainly be damaged by these operations and any areas that are blowing

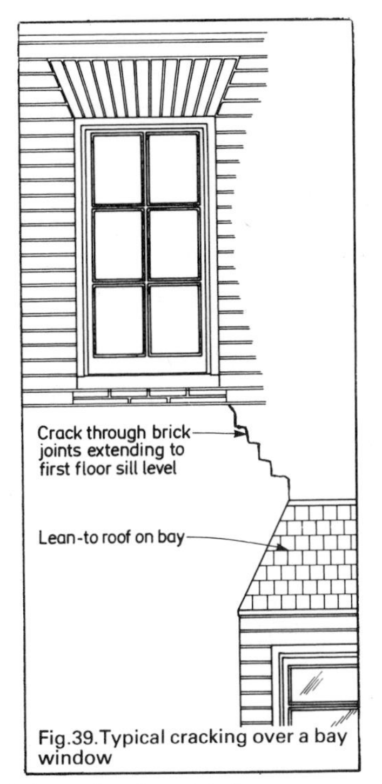

Fig.39. Typical cracking over a bay window

away from the brickwork should be removed and allowances made for making good at a later date.

It is quite common nowadays, when there are defects in a bay window, to remove the whole bay window and install a 'trendy' bow front Georgian window. This does tend to look out of place, being mixed up in a Victorian style home but is nevertheless a very popular improvement. As the new opening may need to be larger, calculations will probably be necessary for submission to the local authority to show the suitability of the beam/RSJ intended to be used.

Even in the case of a straightforward replacement it might be necessary to provide additional permanent support for the new beam to the brickwork at either end. Eighty years ago padstones were often not used and so it is recommended, and the Building Inspector may insist, that concrete padstones are cast and properly built-in to help spread the load.

Another problem that may arise is the difference in depth between the existing bressummer and the new beam. It is highly unlikely that the new beam will need to be deeper than the existing, but it is advisable to oversize the beam should calculations show that a shallower depth would suffice. This would probably eliminate any difficulties in making good.

When the brickwork has been inspected and strengthened as necessary and any padstones inserted, the beam itself can be installed. Half hour fire protection will be necessary and this can be achieved with two sheets of 12.5mm plasterboard with staggered joints and a skim coat of plaster applied afterwards when the internal making good is carried out. Externally, the beam will need to be protected from the elements and pointing/rendering should be made good and repaired to ensure that rainwater cannot penetrate and affect structural stability.

It would also be a good idea, if uncertain about any of the timbers, to invite a timber treatment specialist to inspect and report on the condition of the timbers in the vicinity of the bay window and to estimate for treatment or repair. He will give a guarantee that you, as a general builder, would not be able to give.

When replacing the pitched roof coverings of the bay lean-to you will need to renew the flashings. This will protect the roof timbers from rainwater damage and also alleviate the necessity for repair and maintenance for a number of years.

If there is a crack over the bay window extending through the brick joints, or even through the bricks themselves, up to the first floor windows then it is likely that more substantial repairs will be required. Underpinning may be necessary and the next Project should be consulted. It may be that cracks to the brickwork above the bay window and cracks to first floor sills have caused the deterioration of the bressummer beam in the first place. It is obviously important that all aspects of repair necessary in the vicinity of the bay window are carried out at the same time and that the cause of the defects is rectified along with the symptoms.

When the work is completed, internal making good to any defective or previously hacked off areas of plastering will need to be carried out, along with re-running of any fibrous cornices, etc. The ceiling to the ground floor reception room may be so badly damaged as to be beyond repair. In this case it is recommended that the whole ceiling, laths and plaster are removed and a new plasterboard ceiling erected and skimmed.

Finally, be sure to remove all rubbish and debris from the site, thoroughly clean out the room and hallway and leave the house as clean as you would expect to find your own.

Works to small bay window

1. Replacement of Bressummer Beam
As the text indicates, considerably more work than replacing the beam could be involved depending on the cause and extent of the damage. For the purpose of this Project however we have restricted the costing to the beam alone and consideration must therefore be given to the other factors highlighted before submitting a price.

		£
a)	Allow to move and/or protect furniture and floor coverings in the affected room;	
	Labour: 1 man .25 day	12.56
b)	Remove floorboards to inspect joists prior to shoring;	
	Labour: 1 man .25 day	12.56
c)	Cut away plaster and/or panelling to expose beam and brickwork above and at seatings and clear from site;	
	Labour: 2 men .25 day	25.11
	Plant: skip	5.00
d)	Provide, erect, maintain and remove on completion suitable shores to support floor above and cut out for, insert and support needle to carry brickwork over Bressummer;	
	Labour: 2 men 1.25 days	125.58
	Plant: Acrows, timber etc.	50.00
e)	Carefully cut away brickwork above and at ends of Bressummer to free and remove same;	
	Labour: 4 men .50 day	100.46
	Plant: skip	20.00
f)	Clean out seatings, supply and bed padstones to carry ends of new RSJ;	
	Labour: 2 men .25 day	25.11
	Material: padstones and mortar	18.00
g)	Supply, hoist into position and bed new RSJ (three metres long);	
	Labour: 4 men .33 day	66.81
	Material: RSJ	55.00
h)	Prepare for and make good brickwork above RSJ and at ends including all necessary pinning up to provide a solid support;	
	Labour: 2 men .50day	50.23
	Material: bricks and mortar part salvaged	10.00

i)	Remove all temporary shoring (priced elsewhere) and make good where needle removed including external pointing to match existing;	
	Labour: 1 man .25 day	12.56
	Materials:	3.00
j)	Encase RSJ in two layers of plasterboard on and including suitable treated timber bearers and set in Sirapite;	
	Labour: 2 men one day	100.46
	Material:	25.00
k)	Make good plaster over and at ends of RSJ;	
	Labour: 1 man .33 day	16.58
	Material:	3.00
l)	Make good decorations; it has been assumed the room will be redecorated and form the subject of a separate estimate;	—
m)	Clear up all rubbish and surplus materials and load away; clean out room and replace floor coverings etc;	
	Labour: 2 men .50 day	50.23
	Plant:	5.00
		787.25
n)	Allow for prelims 12½%	98.41
		885.66

Should a temporary screen be required in the common area due allowance should be made for this.

The foregoing labour rates are based on £4.60 per hour for a 39 hour five day week plus 40% for profit and overheads. Should any of these factors differ from those to be used appropriate adjustments should be made accordingly.

Project 21
Works to small bay window - II (bay underpinning)

First discover the reason for the structural defects to a bay window and whether underpinning is really necessary. It is assumed that the works are being carried out to the same small mid-terrace Victorian house as in Project 20 about bressummer beams. When these houses were built, the foundations underneath the bay window were normally extremely shallow and in some cases did not really exist at all.

It is unlikely that the cause of structural defects to the bay will be settlement. The building, being 80 years old or more, will probably have settled many years ago although differential settlement is possible if the bay is on shallower foundations. Subsidence would be more likely and the possibilities of an insurance claim should be investigated. These problems would not suddenly occur; settlement of an old building such as this takes place over quite a number of years and it would be wise to investigate and see whether making good and tying-in has been carried out in the past.

Tree root problems, cracked and defective drains, leaking gullies and rainwater pipes, etc., would all help to cause expansion and shrinkage to a clay subsoil or to wash away a sandy or gravel subsoil. It is recommended that a trial pit be dug to determine depths of foundations in the vicinity of the bay and soil tests carried out as necessary.

Anyone can give a guarantee for the work that they do and many insurance companies, if they are involved, will insist that a 20 year guarantee is given for the works carried out. This does not necessarily mean that the work will need to be carried out by an underpinning specialist although it is recommended that a specialist be involved if the works are major. Other than that, a specialist consulting engineer would need to be employed to design the underpinning and probably carry out the soil tests.

It is also recommended that the drains be tested, as damage to drains would probably be covered under a building insurance policy.

Building Regulations approval will need to be sought and there are small fees that need to be paid to a local authority when an application is made. The structural engineer will need to prepare drawings showing beams and pads, etc., along with any reinforcement design that may be necessary.

It is assumed that the underpinning job is not too complicated and as long as the engineer is employed to supervise the work on site and his designs are accepted by the local authority a small underpinning job will not be beyond the capabilities of a competent general builder.

Specialist plant will be needed, such as a waterpump to clear the pits of any rainwater, as well as a hoist for removing the spoil should this be necessary. Rubbish container skips will also need to be placed in reasonably convenient locations for carting away the excavated materials. If the skips cannot be sited nearby, adjustment should be made on the estimate.

It is important to note that underpinning work should be carried out in stages. If you are unsure about the technicalities or an inexperienced builder it is best to subcontract the work to a specialist firm and just involve yourself with the ancillary builder's work. Underpinning, however, is not a magic or mysterious process. Some knowledge of underpinning would be useful but a good engineer and proper site supervision will ensure that the job is carried out professionally and efficiently.

Most insurance companies will insist that an engineer's report be prepared showing that the progress of the job was satisfactory and that the end result can be guaranteed to perform the job for which it was intended. The client may have problems selling his house without such a report or a guarantee.

Underpinning is basically a way of transferring the loadings imposed on the ground by a building to an acceptably firm subsoil

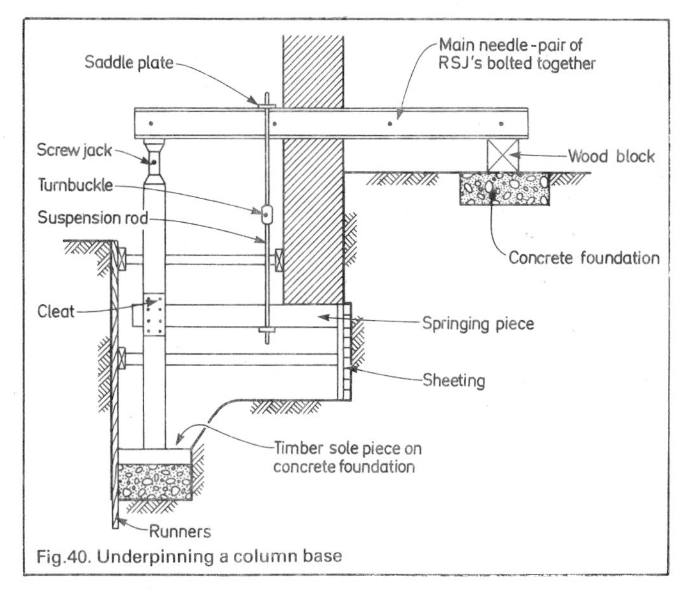

Fig.40. Underpinning a column base

base. Figures 40, 41, and 42 show some similar underpinning details; the principles remain the same for the small job that we are considering. When estimating for the job it will be relatively easy to calculate the amount of concrete required as well as the costs of any reinforcement that may be necessary.

Make allowances for inclement weather as this is probably the main unknown factor that the builder will encounter, and not only will this increase the amount of time required to complete the job but it will also decrease the amount of profit that may be made. One problem that underpinning firms encounter is the need for additional underpinning which can only be discovered as the work proceeds. When underpinning do not make promises to start your next job at any particular time.

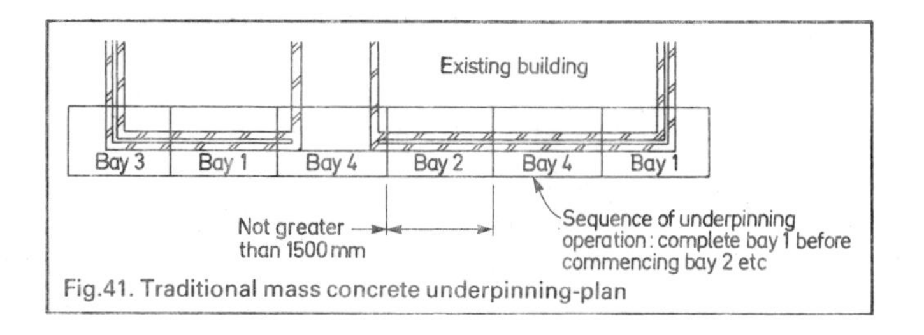

Fig.41. Traditional mass concrete underpinning-plan

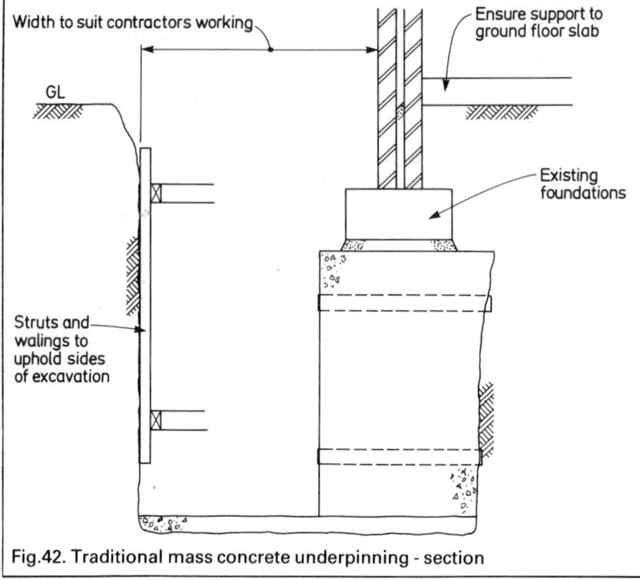

Fig.42. Traditional mass concrete underpinning - section

Making good after underpinning works can often result in greater expenditure than the cost of the work itself. Window sills may need to be recast, or even renewed, and any cracked and defective areas of brickwork will probably need to be cut out and completely rebuilt.

In cavity brickwork, walls may be so badly cracked as to require complete rebuilding and wall ties may have been disturbed or broken and need to be replaced also. Small cracks to brickwork and joints can sometimes be resin-bonded to avoid the necessity for actually cutting out the bricks.

Other areas will need to be repointed and windows eased and adjusted. Floors may have dropped and need to be replaced and all sleeper walls and wallplates replaced. Make sure that damp-proof courses are not adversely affected, and if they are, remember to replace them with all joints thoroughly lapped.

Lack of bearing can sometimes occur to timber suspended floors, resulting in springiness to the floor itself. Sometimes this can be overcome by the addition of a dwarf sleeper wall; in the case of the first floor, more extensive repairs may be necessary.

Make sure that the floors finish level. You are unlikely to encounter any solid concrete floors in the vicinity of the bay window but nevertheless these too should be investigated, should they exist.

In the case of slate damp proof courses these will almost certainly have cracked if subsidence has taken place to the bay and while they may still appear to be intact it is recommended that they be either replaced with a new physical damp proof course or a silicone-injected damp proof course in conjunction with other works.

The roof structure of the bay might be affected and it may have been pulled away from the main building to such an extent that rainwater penetration is taking place or the flashings are disturbed.

It is important to protect the householder's belongings and ensure that things remain weathertight and secure. The client may be a proud gardener but his garden will almost certainly have been wrecked during the process of the job and you should determine whether or not he wishes the making good of his gardens and paths to be covered in the estimate. You might also have had to demolish part of a garden wall and allowance should be made for rebuilding works.

For Building Regulations approval the inspector will wish to see the various stages of the job, as will the engineer for his report. Do not proceed with too much of the job without either one of these interested parties having made their inspections.

Finally the amount of concrete that could be necessary, the mix required as well as the consolidation process will need to be inspected and various tests (such as the concrete slump test) could be asked for.

Remember that if care and planning are used in the initial stages as well as during the course of the job there is no reason why a first class underpinning operation cannot be achieved.

Works to a small Bay Window No. 2 (underpin bay)

For the purpose of this Project we have assumed a bay window overall width of 3.50 metres by 1.00 metre projection from the front of the house.

		£
a)	Clear area around perimeter of bay and take up and clear away any existing paving;	
	Labour: 2 men .25 day	25.11
	Plant: skip, say	10.00
b)	Remove furniture and floor coverings adjacent to bay walls and lift floor boards to gain access to bay wall foundations;	
	Labour: 2 men .25 day	25.11

		£
c)	Excavate pit external to bay to establish depth and nature of existing foundations;	
	Labour: 1 man .16 day	8.39
	NOTE: Investigation of foundations indicates these consist of two brick spreader courses under the 9" bay wall to a total depth of 12" below ground level.	
d)	Excavate to a total depth of 1.00 metre (or such depth as is required by the engineer) in lengths not exceeding 1.00 metre, under the existing foundations and allowing for working space internally and externally (total length approximately 4.40 metres) and cart away surplus spoil;	
	Labour: 2 men (6.6cm) 3 days	301.38
	Plant: skip	40.00
e)	Allow for temporary supports to sides of excavation dependent upon type of subsoil etc.;	
	Labour: 1 man 1 day	50.23
	Material: timber, say	40.00
f)	Carefully cut away overhanging brick spreader courses;	
	Labour: 1 man .50 day	25.11
g)	Concrete filled into trench in short lengths well tamped and consolidated to leave no voids;	
	Labour: 2 men 2.25 days	226.04
	Material: 6.6 c.m (small quantities) concrete	330.00
h)	Carefully pin up under existing brick wall with 1:3 cement and sand well punched in leaving no voids;	
	Labour: 1 man .75 day	37.68
	Material: mortar	3.00
i)	Clear away all rubbish and reinstate flooring internally;	
	Labour: 2 men .75 day	75.35
	Plant: skip	10.00
		1207.40
j)	Allow for prelims 12½%	150.93
		1358.33

NOTE:
Consideration should also be given to such items as reinstatement of paving, flowerbeds etc. as noted in the text, as these items can add considerably to the cost. Due allowance should also be made for all fees and dues chargeable to the work.

The foregoing labour rates are based on £4.60 per hour for a 39 hour five day week plus 40% for profit and overheads. Should any of these factors differ from those to be used appropriate adjustment should be made accordingly.

Project 22
Works to small bay window - III (rebuild bay as existing)

Sometimes, if a bay window is falling away from the main house due to differential settlement or lack of restraint, or if there are other problems that have caused excessive movement of the bay window, it may be preferable to rebuild the whole bay rather than try to effect repairs. Figure 43 shows a typical example.

Allow for the careful removal of the existing bay. It may be necessary to re-use some of the existing materials. Should problems prove to be serious, temporary shoring, as shown in Figures 44 and 45, might be necessary to help support any defective areas of the front elevation. The whole structure will obviously need to

Fig.43. Details of a typical Victorian bay window to a mid-terrace house

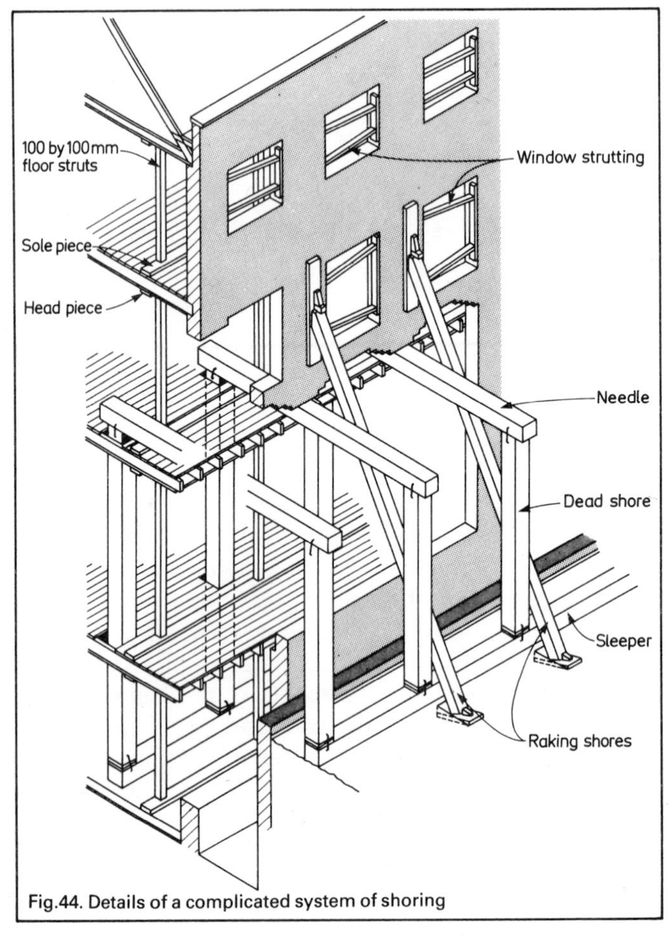

Fig.44. Details of a complicated system of shoring

meet the Building Regulations and the Health and Safety Executive might show an interest.

It is assumed that the building is exactly the same as in Projects 20/21, although if the building is listed an application for Listed Building Consent may be necessary. The local authority will be able to provide further information. The district surveyor's office or the Building Regulations department of the local council will need to be informed of any major works such as these and an application for approval under the various Acts will be necessary.

Foundations will have to be removed and the site prepared for the laying of new foundations. These should still tie in relatively

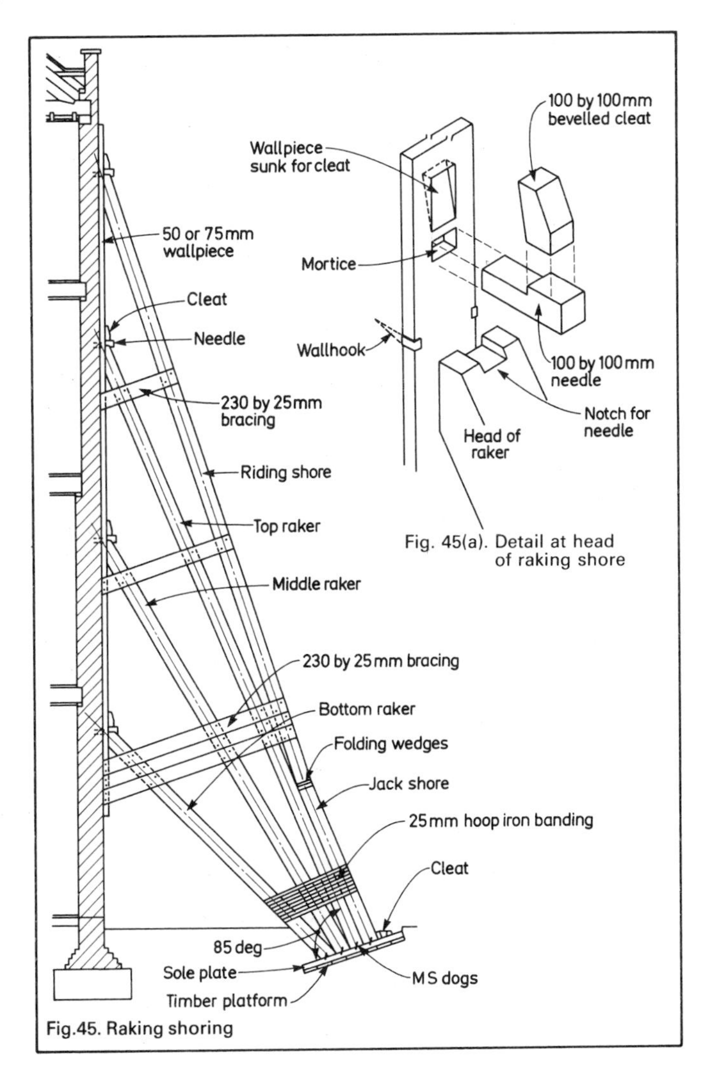

Fig. 45(a). Detail at head of raking shore

Fig.45. Raking shoring

well with the existing foundations. Try to avoid differential depths if at all possible.

Prior to demolishing the existing bay all the services in the vicinity of the bay should be disconnected. The various electricity, gas and water boards should be contacted to determine the whereabouts of these mains.

It is recommended that the existing foundations to the main house be inspected while the opportunity exists and any defects brought to the attention of the client. Drains should also be tested and any repairs or renewals allowed for. It may even be wise to ask the client to carry out the drains test prior to submitting an estimate. No client really likes to see a provisional sum for drain repairs in a quotation.

Any old materials that have been salvaged, such as bricks, slates, windows, etc., should be carefully laid aside and cleaning up and preparation carried out at an early date. It is somewhat unlikely that the client will wish to re-use the existing windows in the bay but purpose-made double-hung sashes can prove to be very expensive and if the windows are in a reasonable condition, or if just minor repairs are necessary, it is as well to let him know of this possibility.

Allow for a new dpc properly tied into, or lapped onto, the existing dpc, wallplates (treated previously) and any other new timbers that may be necessary.

When the brickwork is being rebuilt it would be a good idea to try and match the existing mortar wherever possible but to achieve the same mortar mix will be very difficult. Suggest to the client that the whole front elevation of the house be repointed. This will help to diguise the fact that the bay has been rebuilt, although cleaning down of all the bricks and total redecoration of the front elevation would be the best answer.

It should go without saying that the new work should be tied into the existing properly and any new floor joists, etc., properly built in. The building inspector or district surveyor will want to view the job at various stages and when permission has been granted the relevant office will advise on what procedure to take .

It will be best to renew the timbers to the roof such as rafters, battens, etc., but try where possible to re-use the existing slates laid on treated battens and untearable roofing felt as is common pratice. Flashings between the bay window roof and the main wall should be of Code 4 lead.

The old sill to the bay window will not survive the demolition and consequently a new sill should be cast. It is a good idea to consider a silicone water-repellent coating, such as Ness Synthasil to the whole of the external face of brickwork as many bricks of the Victorian era become porous over a period of time.

External decorations will need to be carried out to the joinery and sills, etc., and complete external redecoration to the front is advisable.

Other making good that should be carried out internally will be very similar to that in Projects 21/22.

Works to a small bay window No. 3 (re-build bay as existing)

For the purpose of this Project we have assumed a bay as described in Project 21 and have also assumed that the front wall above the bay is adequately supported on the existing or replaced bressummer.

		£
a)	Remove furniture and floor coverings from the area adjacent to the bay window and cover with dust sheets etc. for protection;	
	Labour: 1 man .25 day	12.56
b)	Provide and erect a weatherproof screen within the room in the area of the bay to provide full security and adequate protection;	
	Labour: 2 men .50 day	50.23
	Material: Say £250 less credit £125,	125.00
c)	Carefully take down existing bay to ground level, load into skip and cart away. NOTE: Assumed no materials are re-useable Adjustments should be made where appropriate for any salvageable items.	
	Labour: 2 men two days	200.92
	Plant: skip, say	100.00
d)	Excavate for and grub out existing brick and/or concrete foundations, load into skip and cart away;	
	Labour: 2 men one day	100.46
	Plant: Kango	20.00
	Skip, say	50.00
e)	Excavate as necessary to main foundation level or such level as required and cart away surplus spoil to form new bay foundation;	
	Labour: 2 men one day	100.46
	Plant: skip, say	15.00
f)	Concrete in foundations;	
	Labour: 2 men .33 day	33.15
	Material: say one cubic metre	60.00
g)	Brickwork in foundations including bonding in to existing walls with all necessary cutting away and lay and bed d.p.c.;	
	Labour: 2 men two days	200.92
	Material: bricks	58.00
	Sand/cement/d.p.c.	14.00

h)	Brickwork above d.p.c. including forming piers for and building in windows and including cutting, toothing and bonding to existing;	
	Labour: 2 men three days	301.38
	Material: Bricks/mortar	108.66
	Plant: Scaffold, say	30.00
j)	Precast concrete sill and bedding in mortar;	
	Labour: 2 men .33 day	33.15
	Material:	51.00
k)	Precast concrete lintels and building in;	
	Labour: 2 men .33 day	33.15
	Materials:	30.00
l)	Supply and build in 3 No. box sashes to match existing pattern (assumed purpose made);	
	Labour: 2 men .33 day	33.15
	Material:	500.00
m)	Timber to roof including joists, plates and rafters, soffit and fascia;	
	Labour: 2 men one day	100.46
	Material:	52.00
n)	Timbers to floor including joists, plate and floor boards;	
	Labour: 2 men .50 day	50.23
	Material:	54.00
o)	Roof covering including battens, felt, insulation and tiles;	
	Labour: 2 men one day	100.46
	Material:	145.00
p)	Lead flashings to roof including raking out joints and pointing;	
	Labour: 2 men .50 day	50.23
	Material:	30.00
q)	Glaze windows with ordinary quality clear sheet glass with putty and sprigs;	
	Labour: 1 man .50 day	25.11
	Material: glass and putty	70.00
r)	100mm UPVC gutter and 68mm UPVC downpipe including all fittings, brackets etc.;	
	Labour: 2 men .75 day	75.35
	Material:	25.00
s)	External decorations to bay only (assumed if front elevation to be painted this will be subject to a separate quotation);	
	Labour: 1 man 1.50 days	75.35
	Material:	10.00
t)	Plasterboard and set to soffit of bay and render and set in Carlite to internal walls of bay including making good to existing;	
	Labour: 2 men one day	100.46
	Material:	15.00

u)	Skirtings, dado rail, picture rail and window boards to match existing;	
	Labour: 1 man .75 day	37.67
	Material:	35.00
v)	Clear up around outside of bay and make good paving etc. and leave clean and tidy;	
	Labour: 2 men .50 day	50.23
	plant: skip	12.00
w)	Remove internal screen and reinstate floor covering and furniture and leave clean and tidy for tenant;	
	Labour: 2 men .50 day	50.23
	Plant: skip	15.00
		3439.97
x)	Allow for prelims 12½%	430.00
		3869.97

No internal painting has been measured as it is assumed the whole room will be re-decorated and be subject to a separate quotation.

The foregoing labour rates are based on £4.60 per hour for a 39 hour five day week plus 40% for profit and overheads. Should any of these factors differ from those to be used appropriate adjustment should be made accordingly.

Project 23
Re-pointing a wall over a public highway

The usual job on domestic buildings with difficult accessibility is working on the rear of a property that is in the middle of a terrace of houses. This means carting materials or debris through the house, thus increasing the cost of the job as internal protection has to be provided, and most people don't like wheelbarrows running through their houses! You may also be asked to replace the drainage under a house, which causes enormous upheaval to the occupants or, more commonly nowadays, replace or make repairs to an internal soil stack.

Before starting work on the wall contact the local authority which will put you onto its highways department because you will need to erect a scaffold, the local council and various conditions have to be complied with.

Most scaffolding contractors will be familiar with the regulations and requirements although it is important to know that the scaffolding will have to be lit and any protruding edges covered to protect pedestrians.

If the highway is narrow then a temporary pavement will need to be provided over part of the roadway and this will provide further complications.

Such a complex scaffolding system will cost a fair amount of money both to erect and to hire from week to week. If the job is to be done in the winter and could be temporarily stopped by inclement weather, who is responsible for paying the rental charges on the scaffolding? Make sure in your estimate that your client is

aware of the possibility that work could stop and either allow several weeks hire charge when the estimate is prepared, in order to cover such an eventuality, or let the client know that if you cannot complete the work within a couple of weeks, extra hire charges may be added.

When the old pointing is raked out you won't be able to just let it fall on the pavement. It would be difficult to catch the old materials as they drop and so you should allow for some form of tarpaulin or fan to be suspended from the scaffold to catch the debris as shown in Fig.46. Make sure that you clear this debris at the end of the working day. Once again, a competent scaffolding contractor can advise you.

Figures 47 to 50 show other scaffolding details.

The estimating of the job is relatively straightforward as there is only raking out of the pointing, clearance of material, preparation before repointing, such as Unibond, etc., and the repointing itself. You should attempt to match the new pointing with the existing wherever possible, but if the whole wall is being repointed this will not be a problem.

It is recommended that you suggest to your client the carrying out of other works while the scaffolding is in place as the high erection and rental charges will increase the cost of repointing to a very high rate per square metre. A roof overhaul, external redecoration, repairs or renewals to gutterings and fascia boardings may be items that will need to be carried out, and will greatly

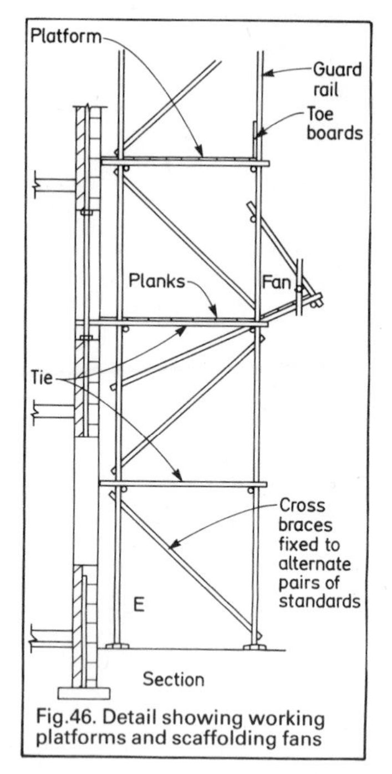

Fig.46. Detail showing working platforms and scaffolding fans

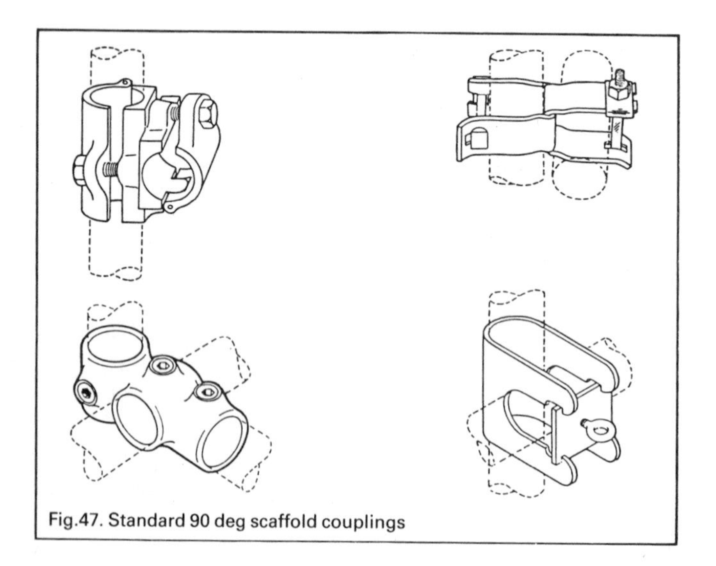

Fig.47. Standard 90 deg scaffold couplings

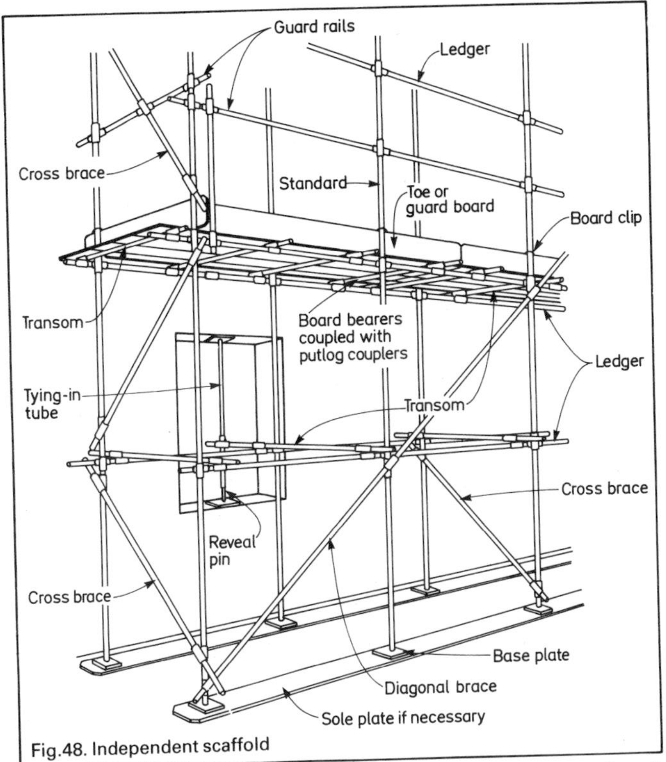

Fig.48. Independent scaffold

reduce the overall cost. It is a false economy to try and only carry out something that may be the bare minimum to render the exterior of the building in a good condition.

To gain access to a scaffold you will need ladders. These should not be left in place when there are no operatives on site not only for safety reasons and to prevent children from climbing the

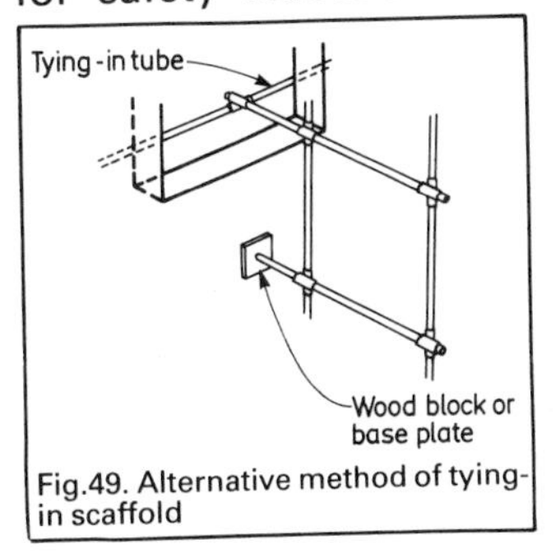

Fig.49. Alternative method of tying-in scaffold

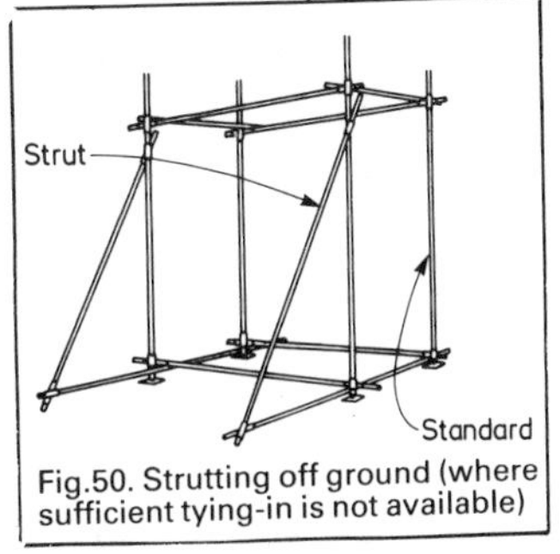

Fig.50. Strutting off ground (where sufficient tying-in is not available)

scaffold, but also to prevent easy access for burglars, especially at first and second floor levels. You might even need to remove the ladders at the end of every working day and take them back to the yard. When using ladders do not place them in a position where pedestrians can walk into them or must walk into the road because of them.

Allow for a hoist to be sited on the scaffold for the lifting and lowering of materials and debris. It may not be possible to site a rubbish skip close to the job and if old materials have to be moved a long distance this will increase the time spent and the cost.

Re-pointing a wall over a Public Highway

For the purpose of this Project we have assumed a building 6.00 metres wide×9.00 metres high with a pavement 4.00 metres wide.

		£
a)	Provide independant scaffold to complete elevation (60 SM) including provision of fan and protective heavy gauge polythene sheets secured to scaffold. Lowest level above pedestrians head height to be close boarded. Clear away on completion;	
	Labour: 3 men 2.50 days	376.73
	Plant: Scaffolding (hire charge)	90.00
b)	Erect plywood on softwood hoarding around lower 3.00 metres of scaffold and provide and maintain warning lights;	
	Labour: 2 men one day	100.46
	Material: (allow 50% credit on return)	135.00
	Plant:	50.00
c)	Rake out joints and re-point in cement and sand to complete elevation (assume window area 16 SM) 44SM;	
	Labour: 2 men three days	301.38
	Material:	25.00
d)	Cart away all rubbish and surplus material;	
	Labour: 1 man .50 day	25.11
	Plant: skip	15.00
e)	Take down and clear away hoarding;	
	Labour: 2 men .50 day	50.23
	Plant: skip	15.00
f)	Transport for scaffold and hoarding material;	50.00
		1233.91
g)	Allow for prelims 12½%	154.24
		1388.15

NOTE: Other costs may arise in respect of replacing defective bricks, making good sills or cornices, replacing air bricks etc. for which due allowance should be made. The estimate should also include for all fees etc. that may be payable.

The foregoing labour rates are based on £4.60 per hour for a 39-hour five day week plus 40% for profit and overheads. Should any of these factors differ from those to be used appropriate adjustment should be made accordingly.

Project 24
Modernising a 1930s kitchen

When the suburbs of London expanded during the 1930s the houses built were considered to be fairly modern although the kitchens, by today's standards, were extremely old fashioned. Many of them only had a worktop, maybe a wall unit and an old Belfast-type sink and wooden draining board.

Your client will probably want to completely refit the kitchen from scratch and you should estimate for the complete removal of everything existing. The cold water mains are likely to rise in the kitchen and will probably be of lead, which has now been proven to be unhygienic. It has been assumed that the electrical installation in the kitchen will also need to be improved and the number of socket outlets increased.

The kitchen should be properly planned. This does not necessarily mean that the owner of the house will have to pay a kitchen-planning consultant but it is important to discuss with your client his requirements and how much he intends to spend on the refurbishment.

There is also likely to be a built-in dresser in the kitchen which might look very nice if the paint was stripped off and the wood exposed, but it may not be practical to keep this dresser, especially in a small kitchen.

When everything is stripped out of the kitchen, the floor should be inspected as it is likely in a property such as this that there will be a timber suspended floor.

When estimating for the kitchen refurbishment it will probably not be possible to determine the extent of any unforseen problems

Old-fashioned sink and wooden draining board.

to the floor, etc., and consequently it is probably better to include a provisional sum for renewals to floor joists, sleeper walls, etc.

In a property dealt with recently, the hot water cylinder, with an electric immersion heater, was located on the kitchen floor in one

Narrow galley type kitchen.

corner of the room. This had been leaking over a number of years and had caused extensive damage to the floor joists, etc., and many of the timbers had to be replaced. You may also find damp problems: a dpc will have to be installed along with the usual renewals of internal plastering, etc.

Underfloor ventilation might also be inadequate and you should determine whether or not this is the case, prior to estimating, and allow for any additional ventilation grilles or airbricks that may be needed.

While floor boards, etc., are being raised you should also check the gas supply to the kitchen and rearrange this or repair it as necessary. Does your client intend using a gas-fired refrigerator and is the position of a gas cooker going to change? It may not be possible to alter the position of the gas supply when all the units have been installed.

The water supply was mentioned above: if the main rises in the kitchen there may not be a stopcock to shut down the supply should the need arise. Many property owners simply rely on the Water Board's stopcock in the road. It is far better to be able to shut down the supply in the house, in the event of an emergency.

It has been assumed that the whole kitchen will need to be replumbed, especially as there is likely to be a washing machine, and possibly a dishwasher, installed and the sink itself is likely to be in a different position. The next Project will discuss plumbing for a washing machine and go into greater depth regarding the problems faced with running supplies and wastes. Should a dishwasher be required, similar problems will be encountered.

Any alterations to the drainage will require approval under the Building Regulations or London Building Acts and thought should be given to the problems of supply and waste when the kitchen is planned.

If a tumble drier is to be installed in the kitchen it is advisable to buy an airventing kit and place the tumble drier adjacent to an external wall so that the venting kit can be installed in the wall itself. This will greatly reduce condensation.

A cooker hood over a hob or free-standing cooker is also a good idea, to reduce condensation in the room. There are two types of cooker hood; one is filled with a granular material to absorb moisture, the other is fitted with a kit to vent it to the outside air, similar to the tumble drier.

When planning the layout of the kitchen with your clients you might find that carrying out some minor structural alterations will greatly improve the overall layout of the new kitchen.

In many of the 1930s terraced houses the kitchen was like a galley, with a door at either end. Many people now prefer to have a kitchen/dining room in the house rather than a living room/dining room. It can be advantageous to block the doors leading in and out of the kitchen and to form an opening into the dining room, ensuring of course that there are doors from the dining room to the rear of the property. This tends to give the layout of the kitchen far more flexibility when it comes to planning and siting the units.

With reference to the services in the kitchen, it may be necessary to remove an old solid fuel boiler or an electric immersion heater, as previously mentioned, and replace this with a new system of water heating. Instantaneous gas-fired water heaters are very popular and efficient. You only heat the water that is required and don't have runs of pipework around the house full of expensively-heated hot water that rapidly becomes cold.

Should the windows, doors, etc., need to be replaced, Projects 1, 2, and 5 will be of assistance. Any damp proofing works found necessary should also be dealt with as discussed in Project 17.

When all the old units have been removed and you are sure that the floor of the kitchen is sound, the conduits and wiring for any new socket outlets or cooker point needed should be installed and

Crosby self-assembly kitchen called Pastelle

the walls replastered or the plastering made good where necessary. As the whole kitchen is being totally refurbished it might be prudent to totally renew the old lath and plaster ceiling with plasterboard sheet, skimmed with a finish coat of plaster.

Your clients may have already decided the type and make of unit that they would like in their new kitchen but you should advise them on other types of unit available and the differences in price and quality. Quite often the kitchen units sold by the larger diy retail outlets are every bit as well made as many of the far more expensive types and often, although not always, the only real difference is the quality of the doors. A unit that is too modern will tend to date very quickly and it is sometimes best to choose a reasonably plain and easy to care for surface.

Ensure that you have an adequate fixing for the wall units and sometimes (especially with internal or stud partitions) it is advisable to fix battens first.

The floor is unlikely to be even and level, consequently you will probably have to adjust the base units to make sure that everything remains plumb and level.

The doors to the kitchen units are usually hung last and will probably need to be aligned.

Worktops are always preformed nowadays and unless there are any really awkward angles or corners they will probably be relatively easy to fix.

Timber suspended floors will have to be covered with hardboard before sheet vinyl floor coverings can be laid. Some people consider it best to lay the hardboard, and sometimes even the sheet vinyl, over the whole floor prior to fitting the kitchen units although it is obviously cheaper to do this when the units are installed. It is probably best to cover the floor entirely with the hardboard prior to fixing the units and also to lay the vinyl but missing out any areas underneath the units, if this will result in a saving. To create a good seal between the floor and the plinths of the units, oversize the vinyl making sure that it runs under the units.

When the units are installed and the work almost complete there will be the usual decorations and making good, and probably some wall tiling and grouting. It is best to allow a few courses of tiles behind the sink unit and above the worktops to act as an easy-to-clean splashback.

Wallpapering and other items of decoration will also be necessary and you may also be asked to install either an extractor fan or some form of permanent ventilation to the room in a further attempt to reduce condensation.

The disposal of rubbish during the course of the job and the possibility of old plaster and units, etc., having to be taken through the house itself should be considered.

Finally, the clients should be made aware that any money they borrow to finance the kitchen refurbishment project will probably be subject to tax relief on the interest paid as this sort of job is classed as home improvement and Government Legislation currently states that tax relief can be claimed in these circumstances.

Kitchen modernisation

For the purpose of this Project we have assumed a typical between wars semi-detached house kitchen where all fitments are to be removed and new plumbing installed

		£
a)	Assuming the Client has emptied all cupboards, carefully remove these including all fixings, battens, bearers etc. and remove from site;	
	Labour: 2 men .50 day	50.23
	Plant: skip, say	25.00
	NOTE: Unless the sink unit can be removed and replaced the same day, temporary washing up and water supply facilities must be provided. Allow for any anticipated costs in this regard.	
b)	Take out all old pipework and fixings and make ready for new;	
	Labour: 1 man .25 day	12.56
	Plant: skip, say	5.00
c)	Run new 15mm copper supply from rising main (assumed rising main is in kitchen) to cold tap position including provision of stop valve at connection of new to old;	
	Labour: 1 man .25 day	12.56
	Materials: Copper pipe, fittings and valve, say	12.00
d)	Run new 15mm copper hot supply to sink position from existing hot supply in kitchen including valve as before;	
	Labour: 1 man .25 day	12.56
	Material: Copper pipe, fittings and valve, say	12.00
e)	Connect cold and hot supply to taps including tap connectors.	
	Labour: 1 man .20 day	10.05
	Material: Tap connectors, say	4.00
f)	Connectors to washing machine and/or dishwasher are dealt with in Project 25.	

g)	Supply only new kitchen fittings. As these vary considerably in price dependent upon quality and number of units it is normal practice to include a Prime Cost Sum to be adjusted when the customer has chosen the required units.	
	For this Project, say	1000.00
h)	Unload, get in, assemble and *Fix Only* all kitchen units and worktops including the provision of all necessary bearers, battens and fixings;	
	Labour: 2 men two days	200.92
	Materials: say	20.00
	NOTE: The above fixing time is entirely dependent upon the number of units and whether they are ready assembled or knocked down for home assembly. Suitable adjustment will need to be made when the exact specification of the units is known.	
i)	Allow a Prime Cost Sum for Stainless steel sink top and taps.	150.00
j)	Unload, get in and *Fix Only* sink top and taps (in prepared cut out);	
	Labour: One man half day	25.11
	Materials: say	3.00
k)	Cut hole through external wall for waste pipe and make good after;	
	Labour: 1 man .50 day	25.11
	Material: say	3.00
l)	Supply and fix new UPVC waste pipe including suitable trap;	
	Labour: 1 man .33 day	16.58
	Material: say	10.00
m)	Make good old waste pipe hole to match existing finishings;	
	Labour: 1 man .25 day	12.56
	Material: say	3.00
n)	Allow for making good all damaged plasterwork etc. (excluding decorations) and leave ready for re-decoration;	
	Labour: 1 man .50 day	25.11
	Material: say	10.00
o)	Clear up all rubbish and surplus material, clean down etc. and cart away surplus from site;	
	Labour: 1 man .25 day	12.56
	Plant: skip, say	5.00
		1677.91
p)	Allow for prelims 12½%	209.74
		1887.65

NOTE:

i) Due to the nature of this type of work careful consideration must be given to the labour times allowed in relation to the type/quality of units and access generally. Also if labour is not fully utilised then an allowance for standing time should be made.

ii) The foregoing labour rates are based on £4.60 per hour for a 39 hour five day week plus 40% for profit and overheads. Should any of these factors differ from those to be used appropriate adjustment should be made accordingly.

Project 25
Installation of washing machine, water supplies, waste and drainage

It is possible to install a washing machine almost anywhere in a house provided certain conditions can be met. There obviously has to be an adequate fall to the waste pipe, to ensure that the waste water will run away easily, without having to run the pipe below ground floor level if at all possible. At first floor level the waste pipe can run between the floor joists if necessary and then connect to a soil stack. At ground floor level it may still be possible to run the waste pipe below the floor but the connection into the drainage system may not comply with the Building Regulations.

The manufacturers of washing machines recommend a minimum height at which the rubber waste pipe from the machine itself should connect. There will be other manufacturers' recommendations that also should be complied with, such as allowing a space of approximately 25 mm at either side of the washing machine, so that the machine has room to vibrate when it enters the spinning programme.

There should also be a trap at the foot of the vertical waste pipe to ensure that no drainage smells enter the house. Although this may be relatively obvious to most people, it is surprising how many wastes to washing machines do not have traps on them; even if the pipe discharges over an open gully grating there can still be a problem with drain smells if the gully becomes blocked.

The waste pipe should be run along the kitchen walls in a position where it is unlikely to be damaged by other kitchen appliances, chairs, even saucepans, etc., and it should be properly fixed to the wall with an adequate number of clips and supports. There should also be rodding eyes for access at all junctions and bends

Should it not be possible for the waste to discharge over an existing gully, a new one will have to be installed. Building Regulations approval, or approval under the London Building Acts, will be necessary and this should be granted before work commences. As mentioned in other Projects, the district surveyor or building inspector will want to inspect the job from time to time and the local authority should be contacted for further information.

Hepworth plastic drain sections are ideal and are widely accepted by the majority of local authorities nowadays, the HepSleve connections save both time and money, especially on such a short drain run as this. You will need to comply with various sections of the Regulations and the drain run will have to be laid to the correct falls. This fall and the positioning of the gully itself wiil be largely governed by the situation of the manhole as you are going to be breaking into an existing drain run.

Make sure that the drains are inspected before backfilling otherwise you will be asked to expose the new drain run again.

When breaking into the existing manhole (if there is one) there are yet more regulations and requirements that will have to be complied with; remember to allow for an easy bend into the drain run with salt-glazed stoneware channelling in the inspection chamber itself.

While estimating for this work it is advisable to inspect the manhole and inform your client of any other repairs that you consider necessary. The rendering to the walls of the manhole may be defective, or the cover and frame may be loose, etc.

Figure 51 shows a typical gully to manhole connection.

Whether or not you need both hot and cold water supplies or just a cold supply depends largely on the type of washing machine that is being installed. Some machines can be filled with either cold or a mixture of hot and cold and others will only take cold. Running a single cold water supply will obviously be easier and cheaper although it might not always be as economical as a mixture.

It is usually cheaper to heat the water by gas than by electricity. If other water consuming appliances in the property are heated by gas the choice is just as difficult as these may use all the water in a

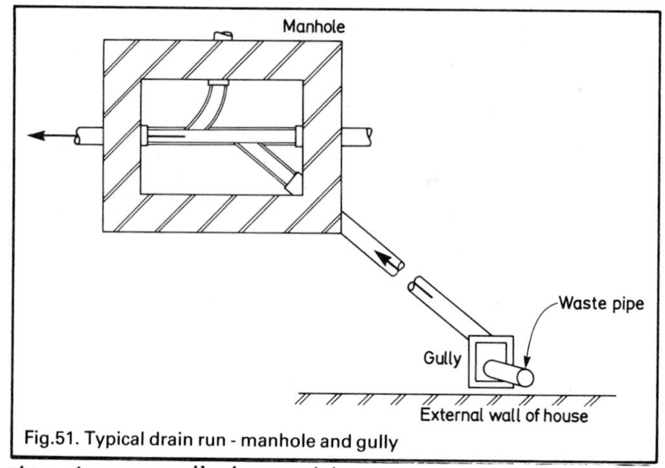

Fig.51. Typical drain run - manhole and gully

hot water storage cylinder and leave none for the machines, or worse still none for a hot bath!

If the machine is likely to be in the opposite corner of the house from the hot water cylinder it is pointless to have 'dead' pipes of hot water lying around unnecessarily.

It is important to make sure that no pipes, either supply or waste, can be accidentally damaged thus causing water leaks. Make sure

Servis Quartz 1000 automatic washing machine. Install machines to manufacturers' recommendations.

that there are adequate clips and brackets to the supply pipes and that even when hidden from view they are installed in a professional and workmanlike manner.

Some people prefer to site their washing machines and tumble driers in the bathroom if there is insufficient room in the kitchen. This can be rather unsightly unless they are tucked away neatly in a well ventilated cupboard but you could experience problems with the water pressure if you are too high up in the house. If you are asked to install the machine at a high level make sure there is adequate pressure first. An electrical fused spur should also be installed in a bathroom, rather than an ordinary switched socket outlet.

It is best to site the washing machine adjacent to the tumble drier and in some instances it is possible to stack the drier on top of the machine. Many larger houses have separate utility rooms which are specifically designed to house such things as washing machines and tumble driers, and the kitchen itself can then be used exclusively for the preparation of food.

Project 24 mentioned that the kitchen should be well planned and this is especially true when considering installing this type of equipment, in order to spare the user's legs.

The shut down valves on the supply pipes are obviously essential as many washing machines are prone to leaking; if the installation is quite simple there are special T-pieces that can be screwed straight onto the supply pipes without having to involve a plumber.

If the property is leasehold don't forget to inform your client that he may have to contact the freeholder for permission to carry out the external works.

Installation of Washing Machine, Water Supplies, Waste and Drainage

For the purpose of this Project we have assumed the washing machine is to be installed in a new utility area with no water and drainage available. Water supply (hot and cold) to be obtained from the adjacent kitchen and a new drain run to be provided to connect to an existing adjacent manhole.

		£
a)	Carefully cut holes through brick partition walls for hot and cold service pipes; Labour: 1 man ⅛th day	6.28

NOTE: This may require kitchen units to be moved temporarily and due allowance should be made for this.

b)	Establish position of hot and cold supplies in kitchen (probably sink risers) remove any obstructions, temporarily turn off supplies, cut into pipes and insert tees, short length of copper and stop valves; reinstate supplies to sink;	
	Labour: 1 man half day	25.11
	Materials: Tees, pipe, valves,	12.00
c)	Supply and fix ½" copper hot and cold supplies from stop valves under sink through previously cut holes in partition to washing machine, position and terminate with hose unions;	
	Labour: 1 man half day	25.11
	Materials: Pipe, clips, fittings etc.	16.00
	NOTE: It may be necessary to empty and move kitchen units to obtain access for the pipes or alternatively the pipes could be made up of short lengths between access points. In both cases this will increase the labour time and possibly also material costs; due allowance must be made accordingly.	
d)	Make good holes through partition to match existing;	
	Labour: 1 man ⅛th day	6.28
	Materials: say	.50
e)	Cut hole through external wall for waste pipe;	
	Labour: 1 man ⅛th day	6.28
f)	Supply and fix UPVC waste from washing machine through wall and terminate over gulley including suitable trap and all fixings;	
	Labour: 1 man half day	25.11
	Material: Pipe, bends, trap, clips etc.	8.00
g)	Connect washing machine to hose union and waste pipe;	
	Labour: 1 man 1/16th day	3.14
h)	Excavate trench for drain including backfill and spread and level surplus (assume 3 linear metres);	
	Labour: 2 men half day	50.23
i)	Break through side of manhole and benching for new 110mm drain connection;	
	Labour: 1 man half day	25.11
j)	Supply and lay pea shingle bed and haunch to 110mm drain;	
	Labour: 1 man ¼ day	12.56
	Material: say	10.00
k)	Supply and lay 110mm UPVC drain including gulley with grid and connection to channel in manhole;	
	Labour: 1 man ⅔rd day	33.15
	Material:	25.00
l)	Make good manhole benching and wall to match existing;	
	Labour 1 man ⅓rd day	16.58
	Material: say	5.00
m)	Test drains	
	Labour: 1 man ¼ day	12.56

n)	Allow to pay all fees in respect of the District Surveyor etc.	
o)	Clear away all surplus materials etc. clean work area and leave ready and working; Labour: 1 man half day	25.11
		349.11
p)	Allow for prelims 12½%	43.64
		392.75

NOTE: the foregoing labour rates are based on £4.60 per hour for a 39 hour five day week plus 40% for profit and overheads. Should any of these factors differ from those to be used appropriate adjustment should be made accordingly.
Consideration should also be given to the possibility of standing time occuring due to non-utilisation of labour.